Dynamics of Atmospheric Flows

Atmospheric Transport and Diffusion Processes

International Series on Advances in Fluid Mechanics

Objectives:

The field of fluid mechanics is rich in exceptional researchers worldwide who have advanced the science and brought a greater technical understanding of the subject to their institutions, colleagues and students.

This book series has been established to bring such advances to the attention of the broad international community. Its aims are achieved by contributions to volumes from leading researchers by invitation only. This is backed by an illustrious Editorial Board for the Series who represent much of the active research in fluid mechanics worldwide.

Volumes in the series cover areas of current interest or active research and will include contributions by leaders in the field.

Topics for the series include: Bio-Fluid Mechanics, Biophysics and Chemical Physics, Boundary Element Methods for Fluids, Experimental & Theoretical Fluid Mechanics, Fluids with Solids in Suspension, Fluid-Structure Interaction, Geophysics, Groundwater Flow, Heat and Mass Transfer, Hydrodynamics, Hydronautics, Magnetohydrodynamics, Marine Engineering, Material Sciences, Meteorology, Ocean Engineering, Physical Oceanography, Potential Flow of Fluids, River and Lakes Hydrodynamics, Slow Viscous Fluids, Stratified Fluids, High Performance Computing in Fluid Mechanics, Tidal Dynamics, Viscous Fluids, and Wave Propagation and Scattering.

Series Editor:

Professor M. Rahman
DalTech, Dalhousie University, PO Box 1000, Halifax,
Nova Scotia, Canada B3J 2X4

Assistant Series Editor:

Dr M.G. Satish
Department of Civil Engineering, DalTech, Dalhousie University,
PO Box 1000, Halifax,
Nova Scotia, Canada B3J 2X4

R.C. Gupta
Department of Mathematics
The National University of Singapore
Singapore 119260

D. Hally
Defence Research Establishment Atlantic
9 Grove Street
PO Box 1012
Dartmouth, Nova Scotia
Canada B2Y 3Z7

C.C. Hsiung
Department of Naval Architecture
Technical University of Nova Scotia
PO Box 1000
Halifax, Nova Scotia
Canada B3J 2X4

D.B. Ingham
Department of Applied Mathematical
Studies
School of Mathematics
The University of Leeds
Leeds LS2 9JT
UK

M. Isaacson
Department of Civil Engineering
University of British Columbia
Vancouver, BC
Canada V6T 1Z4

S. Kim
Department of Chemical Engineering
University of Wisconsin-Madison
1415 Johnson Drive
Madison, Wisconsin 53706-1691
USA

T. Matsui
Department of Architecture
Nagoya University
Furo-cho, Chikusa-ku
Nagoya 464
Japan

T.B. Moodie
Department of Mathematics
University of Alberta
632 Central Academic Building
Edmonton
Canada T6G 2G1

M.A. Noor
Department of Mathematics
King Saud University
PO Box 2455
Riyadh 11451
Saudi Arabia

A.J. Nowak
Institute of Thermal Technology
Technical University of Silesia
44-101 Gliwice
Konarskiego 22
Poland

J. Noye
Department of Applied Mathematics
GPO Box 498
Adelaide
South Australia 5001

M. Ohkusu
Research Institute for Applied Mechanics
Kyushu University
Kasuga-koen 6, Kasuga
Fukuoka
Japan

W. Perrie
Bedford Institute of Oceanography
PO Box 1006
Dartmouth, Nova Scotia
Canada B2Y 4A2

H. Pina
Instituto Superior Tecnico
Av. Rovisco Pais
1096 Lisboa Codex
Portugal

H. Power
Wessex Institute of Technology
Ashurst Lodge, Ashurst
Southampton SO40 7AA
UK

D. Prandle
Proudman Oceanographic Laboratory
Bidston Observatory
Birkenhead
Merseyside L43 7RA
UK

H. Schmitt
Department of Fluid Mechanics Research
Institute of Theoretical Fluid Mechanics
Bunsenstrasse 10
D-3400 Göttingen
Germany

M.P. Singh
Department of Mathematics
Indian Institute of Technology
New Delhi
India

P. Skerget
Faculty of Technical Sciences
University of Maribor
PO Box 224, Smetanova 17
62000 Maribor
Slovenia

P.A. Tyvand
Department of Agricultural Engineering
Agricultural University of Norway
PO Box 5065
N-1432 Ås-NLH
Norway

L.C. Wrobel
Department of Mechanical Engineering
Brunel University
Uxbridge
Middlesex UB8 3PH
UK

M. Zamir
Department of Applied Mathematics
The University of Western Ontario
Western Science Centre
London, Ontario
Canada N6A 5B7

Thanks are due to S. P. Ayra for the figure which appears on the front cover of this book.

Advances in Fluid Mechanics
Volume 18
Series Editor: M. Rahman

Dynamics of Atmospheric Flows

Atmospheric Transport and Diffusion Processes

Editors:

M. P. Singh
University of Alabama in Huntsville, USA

S. Raman
North Carolina State University, USA

Computational Mechanics Publications
Southampton Boston

Series Editor
M. Rahman

Editors
M. P. Singh
Earth System Science Laboratory
University of Alabama in Huntsville
Huntsville
AL 35899, USA

S. Raman
Department of Marine, Earth, and
Atmospheric Sciences
North Carolina State University
Raleigh
NC 27695-8208, USA

Published by

Computational Mechanics Publications
Ashurst Lodge, Ashurst, Southampton, SO40 7AA, UK
Tel: 44(0)1703 293223; Fax: 44(0)1703 292853
Email: CMP@cmp.co.uk
http://www.cmp.co.uk

For USA, Canada and Mexico

Computational Mechanics Inc
25 Bridge Street, Billerica, MA 01821, USA
Tel: 978 667 5841; Fax: 978 667 7582
Email:cmina@netcom.com

British Library Cataloguing-in-Publication Data

A Catalogue record for this book is available from the British Library

ISBN 1853124273   Computational Mechanics Publications, Southampton

ISSN 1353 808X   Series

Library of Congress Catalog Card Number 97-80259

CONTENTS

Preface   vii

Chapter 1
Unstable and Convective Boundary Layers   1
S P Ayra

Chapter 2
Turbulence and dispersion in the stable   39
atmospheric boundary layer
Section 2A
K. Shankar Rao & Carmen J Nappo

Turbulence and dispersion in the stable   93
atmospheric boundary layer
Section 2B
R.T. McNider, M.P. Singh, & S.Gupta

Chapter 3
Urban Air Pollution   131
M P Singh, J Shah and H B Hassan

Chapter 4
Mesoscale atmospheric transport and diffusion processes   155
Z Boybeyi & S Raman

Chapter 5
Leaky containment vessels of air: A Lagrangian-mean   195
approach to the stratospheric tracer transport
Noboru Nakamura

PREFACE

This monograph covers topics on atmospheric transport and diffusion processes with particular emphasis on the Atmospheric Boundary Layer (ABL). The ABL is the region in which the atmosphere experiences surface effects through vertical exchanges of momentum, heat and moisture. This is the part of the atmosphere in which most air pollutants are released and transported by mean winds and dispersed by turbulent eddy motions. With rapid industrialization taking place all over the world resulting in considerable deterioration of the air quality, the importance of study of this region cannot be over emphasised. Although there has been extensive study of the lowest few hundreds of meters of the atmosphere, there is no rigorous theory of turbulence, especially for the non-uniform conditions expected in the atmosphere. For these reasons no fully validated theory exists and any calculation is expected to be subject to considerable uncertainty.

Air pollution, both as a field for scientific investigation and as a factor in our daily lives, is becoming increasingly important. The trend toward urbanization and greater industrial concentration has led, among other things, to congestion in residential areas and heavier use of city highways. These, in turn, have resulted in more widespread contamination of our atmosphere. If the trend continues, the problem of air pollution is likely to become even more severe in the future. The public and industry are aware of the situation, and much has been done in the way of abatement and control.

In dealing with a problem so complex as air pollution, the cooperation of experts in many different areas is required. Each of the contributors to this monograph was chosen because he / she could speak authoritatively on one or more aspects of the discipline, and each is regarded as responsible for the material in his / her section. Each chapter is independent and treats a single part of the broad subject.

The first chapter gives the overview of our current understanding of the ABL under unstable and convective conditions. Commonly used theoretical and numerical models of the ABL are described and the emphasis is on basic concepts used in the formulations rather than on the details of model parametrizations and results.

The second chapter on the stable ABL has been divided into two sections. Section A deals with the turbulent structure of the idealized stable ABL described in terms of similarity theories and observations from field experiments. The current approaches to the modeling of turbulence and dispersion in the stable boundary layer are outlined. Several complicating factors, such as radiative transfer, gravity waves, intermittency, and measurement problems are discussed. Section B treats two important aspects relevant in the context of stable boundary layer. Techniques of nonlinear dynamics and stability analysis have been used to show the existence of multi-valued solutions for certain values of external parameters which have strong implications for the predictability of the stable boundary layer. The second aspect deals with influence of vertical shear in the horizontal wind produced by diurnal and/or inertial oscillations that could maintain plume growth rates consistent with the observations. Radar profiler technology has been used to carry out a time series analysis of the energy spectra. A recently conducted air pollution field study is used to demonstrate the utility of this analysis.

The third chapter deals with urban air pollution. Health effects and economic impact are described. Summary of pollution assessment methods and the sources of pollution are outlined. Implications of dispersion and transport of pollutants affecting the air quality are presented. Recent approaches to air quality management and related action plans are discussed.

The fourth chapter deals with mesoscale atmospheric processes which are important for the short and long range transport and diffusion of pollutants. The understanding of mesoscale atmospheric transport and diffusion processes are still incomplete. Recent field experiments and numerical studies have enhanced our understanding in this area and numerical models with higher resolution and better representation of land surface processes have also contributed to our increased knowledge in this topic.

The last chapter deals with a diagnostic formalism which is developed for studying the Lagrangian-mean motion of a tracer in the Earth's stratosphere. The formalism is applied to the nitrous oxide field simulated by the Geophysical Fluid Dynamics Laboratory general circulation model to illustrate the natural variabilities of the tracer transport in the stratosphere.

The subject material of this book will be of interest to researchers as well as students intending to take up studies in the field of Atmospheric Boundary Layer and Air Pollution Meteorology. The third chapter will be of interest to planners, administrators and scientists working in the regulatory agencies.

Sethu Raman and M.P. Singh
Editors

Chapter 1
Unstable and convective boundary layers

S.P. Ayra
Department of Marine, Earth and Atmospheric Sciences
North Carolina State University
Raleigh, NC 27695-8208, USA

Abstract

An overview of our current understanding of the atmospheric boundary layer (ABL) under unstable and convective conditions is given. The emphasis is on boundary layers over flat and homogeneous surfaces and during fair-weather conditions. Following a brief introduction to the definitions and concepts, simplified equations of conservation of mean thermodynamic energy, mass, and momentum are presented. The observed mean flow and turbulence structure (variances and covariances) of unstable and convective boundary layers is discussed in the framework of similarity theories and scaling. The ABL is divided into the surface layer, the mixed layer and the transition layer for this purpose. Finally, commonly used theoretical and numerical models of the ABL under unstable and convective conditions are described. The emphasis is on the basic concepts used in their formulation rather than on the details of model parameterizations and results.

1 Introduction

The lowest portion of the atmosphere which is directly affected by the earth's surface over short time scales of the order of an hour to a day is called the atmospheric boundary layer (ABL) or the planetary boundary layer (PBL). The PBL forms as a consequence of small-scale interactions between the atmosphere and the underlying land or water surface. Significant exchanges of momentum, heat and mass between the atmosphere and the earth's surface occur through the PBL. Consequently, flow and thermodynamic variables, such as wind velocity, temperature, and specific humidity, vary continuously throughout the boundary layer from their values at the surface to those of the free atmosphere at the top of the PBL. The dominant mechanism for the transfer of momentum, heat, and mass is turbulent mixing or exchange. Molecular transfer is important only in an extremely thin (order of 1 mm) molecular sublayer interfacing with the surface, but it is primarily responsible for the sharpest variations in the meteorological variables in the interfacial layer.

Air being a viscous fluid, the relative velocity must be zero at the surface. Consequently, wind speed increases with height, at least in the lower part of the PBL, and momentum is transferred downward to the surface (the earth's surface is the ultimate sink of atmospheric momentum). Wind direction also varies with height due to the earth's rotation around its axis.

Due to the diurnal radiative heating and cooling of the surface, the surface is often warmer or cooler than the air above. Here, we are primarily concerned with the unstable and convective PBLs above warmer surfaces in which there is an upward transfer of heat. The resulting convective motions supplement and often dominate the shear-generated turbulence with increasing height above the surface. Turbulence is the most conspicuous and important property of the PBL in general and the convective boundary layer (CBL) in particular. It is also the largest impediment to its understanding, modeling, and prediction. Turbulence represents the apparently chaotic nature of many fluid flows, which is manifested in the form of highly irregular, almost random, temporal and spatial fluctuations in velocity, temperature and scalar concentrations around their mean values.

Similar to that of any turbulent flow, our understanding of the PBL comes in large part from observations and experiments. Observations indicate that the physical and thermodynamical properties of the underlying surface, in conjunction with the dynamics and thermodynamics of the lower atmosphere (troposphere), determine the PBL structure including its depth, wind, temperature and humidity distributions, turbulence and diffusion characteristics, and energy dissipation.

The observed characteristics of the unstable and convective PBLs are often described in the framework of surface layer and mixed layer similarity theories (Caughey, 1982; Kaimal and Finnegan, 1994). These and other theories and models of the CBL will be reviewed here. But first we start with some of the fundamental conservation equations of thermodynamic energy, mass, and momentum, which form the basis of PBL theories and models.

2 Conservation equations

2.1 Conservation of thermodynamic energy

The ultimate source of the thermodynamic energy and motions (kinetic energy) in the atmosphere is the sun. Only a small portion of the solar radiation is directly intercepted by the atmosphere, primarily by clouds and aerosols. Most of it reaches the earth's surface and is partly absorbed in a thin layer of the subsurface materials, surface vegetation and other surface roughness elements and is partly reflected back, depending on the surface albedo. In the daytime, there is usually a net radiation flux, R_N, directed toward the surface. This radiative energy is redistributed in different ways.

The exchange of energy between the surface and the atmosphere is primarily through turbulent motions in the lower part of the PBL, called the surface layer. A part of it is in the form of direct or sensible heat flux, H, which is directed upward from the warmer surface during daytime. The other form of turbulent transfer is the latent heat of evaporation, H_L, from the surface. A significant portion of the net radiation is also converted into the ground heat flux, H_G, which is transmitted to the subsurface medium and is responsible for the warming of the submedium to a depth of less than 1 m in soils and more than 10 m in water on a diurnal basis.

The above mentioned partition of net radiation between different components and forms of energy fluxes is usually expressed in the form of the surface energy budget equation (Oke, 1987; Arya, 1988)

$$R_N = H + H_L + H_G \qquad (1)$$

Equation (1) is strictly valid only for an 'ideal' surface which is flat, bare and opaque to radiation. An 'ideal' surface is also assumed to be a plane of infinitesimal thickness and mass, with no heat storage capacity.

For a finite-thickness interfacial layer, such as a vegetative canopy or an urban canopy, a more appropriate energy budget equation is

$$R_N = H + H_L + H_G + \Delta H_s \qquad (2)$$

in which the storage term, ΔH_s, represents the rate of change of internal energy of the layer per unit horizontal area, which may be expressed as

$$\Delta H_s = \int_0^D \frac{\partial (\rho\, cT)}{\partial t}\, dz \qquad (3)$$

Here, ρc is the heat capacity, T is the absolute temperature, and D is the layer thickness or depth. Note that ΔH_s is positive when the layer is warming with time and vice-versa.

The energy balance equation (1) or (2) tells us how the net radiation received at the surface or at the top of a surface canopy might be partitioned between the various other forms of energy exchanges with the atmosphere and the subsurface medium. For the typical daytime unstable and convective conditions over land and water surfaces, the sum of sensible and latent heat fluxes is a substantial part of net radiation and the former can be estimated from the measurements or estimates of R_N and H_G or ΔH_s. Further separation of the total energy flux to the atmosphere into sensible and latent heat fluxes can be made if their ratio, called the Bowen ratio, $B = H/H_L$, can be estimated independently. Both the fluxes show strong diurnal variations with their maximum values around midday.

Other forms of thermodynamic energy equations are (see e.g., Arya, 1988; Stull, 1988; Garratt, 1992):

$$\frac{\partial}{\partial t}\left(\rho_s c_s T\right) = -\frac{\partial H}{\partial z} \qquad (4)$$

for the subsurface medium (e.g., soil) in which ρ_s and c_s are the density and specific heat ($\rho_s\, c_s$ is called the heat capacity) of the submedium, and

$$\frac{\partial}{\partial t}\left(\rho c_p \overline{T}\right) = -\frac{\partial H}{\partial z} \qquad (5)$$

for the PBL, where ρ is the air density, c_p is the specific heat at constant pressure, and $\overline{T}$ is the mean air temperature. The above equations are strictly valid for only horizontally-homogeneous subsurface layer and the PBL, because only the vertical heat transfer was considered in deriving these equations from the principle of conservation of energy.

2.2 Conservation of mass

The conservation of mass of water vapor in the horizontally-homogeneous, fair-weather PBL leads to the rate equation for the mean specific humidity $\overline{q}$

$$\frac{\partial \overline{q}}{\partial t} = -\frac{1}{\rho}\frac{\partial E}{\partial z} \qquad (6)$$

where E is the vertical water vapor flux. Equation (6) expresses the obvious fact that the moistening of air in the PBL is essentially due to the convergence of water vapor flux. Here, we are ignoring the effects of condensation or evaporation processes involving fog and clouds.

The addition of water vapor may significantly change the density of air, particularly its variation with height or density stratification, and must be considered in the thermodynamics as well as the dynamics of the PBL. Its effect on density and other thermodynamic variables is expressed through the equation of state for the moist air

$$P = R \rho T_V \tag{7}$$

where P is pressure, ρ is density, R is the specific gas constant for dry air, and T_V is the virtual temperature defined as

$$T_V = T (1 + 0.61 \, q) \tag{8}$$

It is the temperature dry air would have if its pressure and density were equal to those of moist air. With increasing specific humidity, T_V increases and ρ decreases.

Similarly, the conservation of mass of any scalar in the PBL with an infinite area source near the ground or at the top of the PBL yields

$$\frac{\partial \bar{c}}{\partial t} = - \frac{\partial F_c}{\partial z} \tag{9}$$

where $\bar{c}$ is the mean concentration and F_c is the vertical flux of the scalar. Mean concentrations of trace gases, such as carbon dioxide, methane, nitrogen oxides, etc., are found to be too small to significantly affect the air density and, hence, thermodynamics and dynamics of the PBL.

2.3 Conservation of momentum

Considerations of the conservation of momentum in an elementary volume of fluid yield the Navier-Stokes equations of motion. Upon averaging, the equations for mean horizontal motion in the horizontally-homogeneous PBL are given as

$$\frac{\partial \bar{u}}{\partial t} - f\left(\bar{v} - v_g\right) = \frac{1}{\rho} \frac{\partial \tau_{zx}}{\partial z} \tag{10}$$

$$\frac{\partial \bar{v}}{\partial t} + f\left(\bar{u} - u_g\right) = \frac{1}{\rho} \frac{\partial \tau_{zy}}{\partial z} \tag{11}$$

Here, $\bar{u}$ and $\bar{v}$ are the mean velocity components in the x- and y-directions, respectively, u_g and v_g are the corresponding geostrophic wind components, f is the Coriolis parameter, and τ_{zx} and τ_{zy} are the horizontal shear stresses in x- and y-directions, respectively.

The various fluxes and stresses in the above conservation equations can be expressed almost entirely in terms of covariances involving turbulent fluctuations of velocity and scalars, as molecular contributions to them are negligibly small in large Reynolds-number flows such as the PBL. In particular, outside any interfacial molecular sublayer,

$$H = \rho\, c_p\, \overline{w'T'}; \quad E = \rho\, \overline{q'w'} \tag{12}$$

$$\tau_{zx} = -\rho\, \overline{u'w'}; \quad \tau_{zy} = -\rho\, \overline{v'w'} \tag{13}$$

in which we have used Reynolds' decomposition in expressing an instantaneous variable as a sum of its mean (denoted by overbar) and fluctuating (denoted by prime) parts. The above covariances represent the vertical turbulent fluxes of heat, water vapor and momentum, which represent the important turbulent exchange processes occurring throughout the PBL. The presence of covariances in the Reynolds-averaged equations for the conservation of heat, mass, and momentum makes these an unclosed set which cannot be solved without introducing some hypothetical closure models. The original set of instantaneous conservation equations (not given here) is closed, but is also not possible to solve numerically for large Reynolds-number flows such as the PBL. Some of the commonly used modeling approaches will be discussed in a later section. First we describe the observed mean and turbulence structure of the PBL under unstable and convective conditions.

3 Similarity scaling and observed structure

For the horizontally-homogeneous PBL over a flat and uniform surface, the observed mean flow, temperature and humidity distributions, as well as turbulence structure are best represented in an appropriate similarity theory framework. This makes the results more general and independent of particular sites and time periods of measurements. A similarity theory, based on dimensional analysis, provides the appropriate scales for normalizing the variables into dimensionless groups or similarity parameters and organizing experimental data in the most efficient manner to empirically derive the predicted similarity relations.

For similarity considerations, the unstable and convective atmospheric boundary layers are generally divided into three layers, viz., the surface layer, the mixed layer, and the transition layer or the entrainment layer, as shown in Fig. 1.

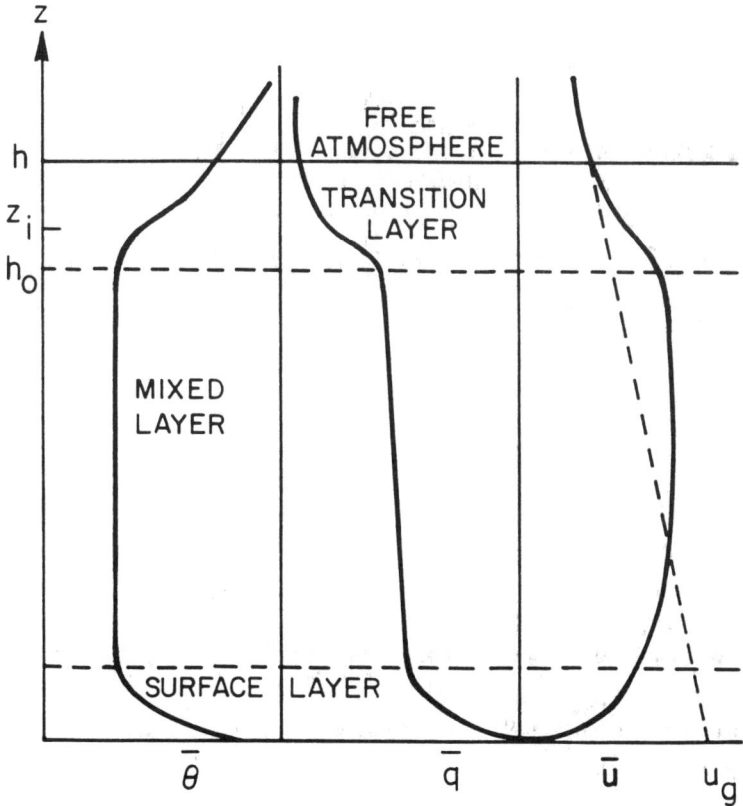

Figure 1: Schematic of the vertical distributions of mean potential temperature $(\overline{\theta})$, specific humidity $(\overline{q})$, and velocity component $(\overline{u})$, showing the three-layer structure of the convective boundary layer.

The surface layer is the lowest 5-10% of the PBL in which significant variations (gradients) of wind speed, potential temperature and specific humidity with height are observed to occur, but mean wind direction remains nearly constant with height. The height above the surface constitutes an important length scale in this layer. Immediately above the shallow surface layer lies the much deeper mixed layer, which extends almost up to the base of the capping inversion. It is characterized by a uniform potential or virtual potential temperature and nearly uniform winds, water-vapor mixing ratios and other scalar concentrations. The mixed-layer depth or the height of inversion base z_i, emerges as the controlling length scale for this middle layer. The mixed layer is capped by a stably stratified transition layer in which turbulence is suppressed with increasing height and completely vanishes at the top of the PBL. This layer is characterized by intermittently penetrating thermals from below and entrainment of the nonturbulent free-atmospheric air from above, particularly during the rapidly growing period. The transition layer typically extends from 0.9 z_i to 1.2 z_i, although both the lower and the upper boundaries can vary over a much wider range, e.g., from 0.5 z_i to 1.5 z_i for the rapidly entraining and evolving CBL.

3.1 Monin-Obukhov similarity theory and scaling

In the lower 10% of the PBL, called the atmospheric surface layer, the mean wind direction and the vertical fluxes of heat, water vapor, and momentum remain nearly constant with height. For the frequently encountered stratified surface layer, Monin and Obukhov (1954) proposed a similarity hypothesis that mean gradients and turbulence characteristics depend only on the height z, the surface stress or momentum flux τ_0/ρ, the virtual heat flux $H_v/\rho\ c_p$, and the buoyancy variable g/T_{vo}. The virtual heat flux represents the covariance $\overline{w'\theta'_v}$ between vertical velocity and virtual temperature fluctuations, which is related to sensible and latent heat fluxes as (Lumley and Panofsky, 1964)

$$H_v = H + \left(0.61\ c_p\ \overline{\theta}\ /\ L_e\right) H_L \simeq H \left(1\ +\ 0.07\ B^{-1}\right) \qquad (14)$$

Thus, the latent heat flux due to evaporation ($H_L = L_e E$) makes a significant contribution to H_v only when the Bowen ratio is much less than one, which generally occurs over extensive water surfaces.

The appropriate scaling parameters for the M-O similarity theory are:

$$u_* = \left(\tau_o / \rho\right)^{1/2}; \quad \theta_* = - H / \rho \, c_p \, u_*$$

$$q_* = - E / \rho \, u_*; \quad L = - \rho \, c_p \, T_{vo} \, u_*^3 / \text{kg} \, H_v \tag{15}$$

where u_* is the friction velocity, θ_* and q_* are the corresponding temperature and specific humidity scales, and L is Obukhov's (1946) buoyancy length scale. Since the height z above the surface is another length scale in the surface layer, its ratio with the Obukhov length, z/L, forms the dimensionless similarity parameter of the M-O similarity theory. This parameter is also a good measure of buoyancy or stability effects in the stratified surface layer and is commonly referred to as the M-O stability parameter.

The similarity theory predicts that dependent variables, such as mean velocity, temperature and humidity gradients, variances of turbulent fluctuations, etc., when scaled (normalized) by appropriate combinations of z, u_*, θ_*, and q_*, must be unique functions of $\zeta = z/L$ only. The most widely used M-O similarity relations are:

$$\frac{kz}{u_*} \frac{\partial \bar{u}}{\partial z} = \phi_m (\zeta); \quad \frac{kz}{\theta_*} \frac{\partial \bar{\theta}}{\partial z} = \phi_h (\zeta); \quad \frac{kz}{q_*} \frac{\partial \bar{q}}{\partial z} = \phi_w (\zeta) \tag{16}$$

in which the dimensionless similarity functions $\phi_m (\zeta)$, $\phi_h (\zeta)$, etc., have been evaluated empirically from micrometeorological experiments over flat and uniform sites (Zilitinkevich and Chalikov, 1968; Dyer, 1967, 1974; Dyer and Hicks, 1970; Businger et al., 1971; Hicks, 1976; Hogstrom, 1982, 1988; Garratt and Hicks, 1990). There appears to be a general agreement that under unstable conditions in the presence of finite, steady winds, the similarity functions are of the form (Arya, 1988; Garratt, 1992)

$$\phi_m = \left(1 - \gamma_1 \zeta\right)^{-1/4} \tag{17}$$

$$\phi_h = \phi_h (0) \left(1 - \gamma_2 \zeta\right)^{-1/2} \tag{18}$$

in which the empirical constants γ_1, γ_2 and $\phi_h (0)$ are estimated from experimental data. There has been no general consensus, however, on the best values of the above constants; the values reported in the literature vary over fairly wide ranges (Hogstrom, 1988):

$$\gamma_1 = 15\text{-}28; \quad \gamma_2 = 9\text{-}16; \quad \phi_h(0) = 0.74\text{-}1$$

Large differences in empirical estimates of similarity functions ϕ_m and ϕ_h by different investigators using experimental data from similarly ideal and flat sites have been attributed to measurement errors in the estimates of mean gradients and fluxes (Yaglom, 1977; Hogstrom, 1988). In particular, the errors due to flow distortion by instruments and mounting towers or masts, inadequate instrument response to turbulent fluctuations, infrequent calibration, and lack of proper probe alignment are suspected to be present in many micrometeorological measurements of fluxes and gradients. For example, flow-distortion errors have been blamed (Wieringa, 1980) for the gross underestimation of the von Karman constant k and $\phi_h(0)$ from the Kansas experiment by Businger *et al.* (1971). More recent studies have confirmed that $k = 0.40 \pm 0.01$ and $\phi_h(0) = 0.95 \pm 0.04$ (Kondo and Sato, 1982; Hogstrom, 1985, 1988). The latter value is not much different from the value of $\phi_h(0) = 1$ assumed or estimated in many earlier studies. Note that this implies an approximate equality between the eddy diffusivities of heat and momentum in near-neutral $(\zeta \to 0)$ conditions. With $\phi_h(0) = 1$, the more widely used Businger-Dyer relations (Arya, 1988; Garratt, 1992)

$$\phi_h = \phi_m^2 \left(1\text{-}16\ \zeta\right)^{-1/2} \tag{19}$$

imply that $\gamma_1 \approx \gamma_2 \approx 16$ and $\mathrm{Ri} = \zeta$,

where Ri is the gradient Richardson number

$$\mathrm{Ri} = \frac{g}{T_{vo}} \frac{\partial \overline{\theta}_v}{\partial z} \Big/ \left(\frac{\partial \overline{u}}{\partial z}\right)^2 = \zeta\, \phi_h / \phi_m^2 \tag{20}$$

However, the more recent evaluations of the M-O similarity functions by Hogstrom (1988) imply different values of $\gamma_1 \simeq 19.3$ and $\gamma_2 \simeq 12$ in Eqs. (17) and (18).

From the limited experimental evaluations of dimensionless specific humidity gradient function $\phi_w(\zeta)$, it appears that in the unstable surface layer with upward transfers of heat and water vapor,

$$\phi_w(\zeta) \approx \phi_h(\zeta) \tag{21}$$

which implies an approximate equality of the eddy diffusivities of heat and water vapor (Dyer, 1967; Paulson *et al.*, 1972). When water vapor and heat are transferred in different directions, however, as in the case of warm air advection over a wet surface, the diffusivities of heat and water vapor are observed to be different (see e.g., Verma *et al.*, 1978). Under such conditions, $\phi_w(\zeta)$ might be expected to be different from $\phi_h(\zeta)$, but reliable estimates of the former are lacking.

The normalized profiles of velocity and potential temperature can be obtained by integrating Eqs. (17) and (18) with respect to z, i.e.,

$$\bar{u}/u_* = 2.5\left[\ln\left(z/z_0\right) - \psi_m(\zeta)\right] \tag{22}$$

$$\left(\bar{\theta} - \bar{\theta}_0\right)/\theta_* = 2.5\left[\ln\left(z/z_0\right) - \psi_h(\zeta)\right] \tag{23}$$

in which z_0 is the roughness length, $\bar{\theta}_0$ is the potential temperature at $z = z_0$, and $\psi_m(\zeta)$ and $\psi_h(\zeta)$ are uniquely related to $\phi_m(\zeta)$ and $\phi_h(\zeta)$, respectively (see e.g., Panofsky and Dutton, 1984; Arya, 1988). The above profile relations may be considered as simple modifications of the log-law in which both the stability-dependent functions increase with increasing instability. Consequently, the profile curvature also increases with increasing instability.

The applicability of the Monin-Obukhov similarity scaling to turbulence in the unstable surface layer is found to be limited to the statistics of vertical velocity and scalar fluctuations. Some measurements of the standard deviation of vertical velocity σ_w in the atmospheric surface layer over land and sea surfaces are shown in Fig. 2, using the M-O similarity framework. These can be represented by the empirical similarity relation (Panofsky *et al.*, 1977)

$$\sigma_w/u_* = 1.3\left(1 - 3\,\zeta\right)^{1/3} \tag{24}$$

Similarly, the spectrum and the large-eddy scale of vertical velocity show systematic variations with the M-O similarity parameter (Gurvich, 1960; Kaimal *et al.*, 1972; Monin and Yaglom, 1975). Standard deviations of scalar fluctuations also generally follow the M-O similarity scaling (Monin and Yaglom, 1975; Bradley and Antonia, 1979). But, the horizontal

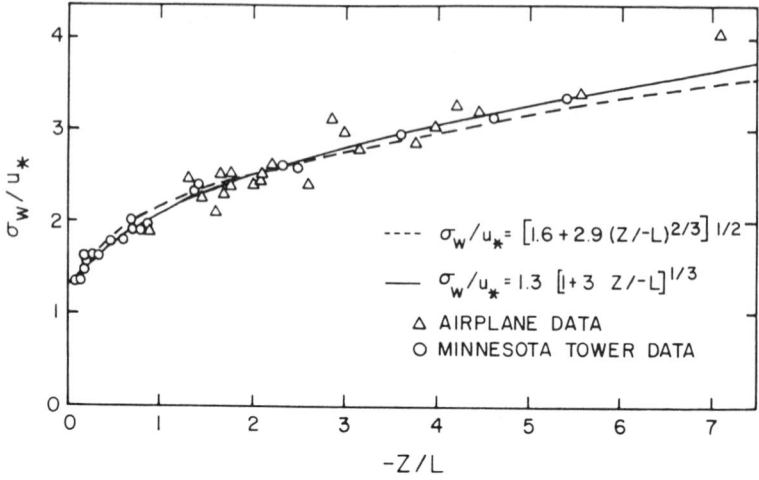

Figure 2: Normalized standard deviation of vertical velocity (σ_w / u_*) as a function of z/L in the unstable surface layer (from Panofsky *et al.*, 1977).

velocity fluctuations in unstable and convective conditions appear to be more affected by large-eddy convective circulations spanning the whole depth of the PBL and do not follow the M-O similarity scaling (Panofsky *et al.*, 1977).

3.2 Local free convection similarity scaling

Another similarity theory of limited validity to only vertical velocity and scalar fields in the convective surface layer is based on Obukhov's free convection similarity hypothesis, according to which u_* is considered irrelevant and only z, g/T_0 and $H_0/\rho c_p$ are considered as independent variables, particularly when $z > -L$. These lead to the following local free-convection scales:

length: z

velocity: $$u_f = \left(\frac{g}{T_0} \frac{H_o}{\rho c_p} z \right)^{1/3} \qquad (25)$$

temperature: $$\theta_f = \left(\frac{H_o}{\rho c_p} \right)^{2/3} \left(\frac{g}{T_0} z \right)^{-1/3}$$

In order to account for the buoyancy effects of water vapor flux, $g\, H_o/T_0$ is replaced by $g\, H_v/T_v$ in the definition of u_f. The important predictions of the local free convection similarity theory are:

$$\frac{\partial \bar{\theta}}{\partial z} = C \left(\frac{H_o}{\rho c_p} \right)^{2/3} \left(\frac{g}{T_0} \right)^{-1/3} z^{-4/3} \qquad (26)$$

$$\sigma_w = C_w \left(\frac{H_v}{\rho c_p} \right)^{1/3} \left(\frac{g}{T_v} \right)^{1/3} z^{1/3} \qquad (27)$$

$$\sigma_\theta = C_\theta \left(\frac{H_o}{\rho c_p} \right)^{2/3} \left(\frac{g}{T_0} \right)^{-1/3} z^{-1/3} \qquad (28)$$

in which C, C_w and C_θ are empirical constants.

There is considerable observational and experimental data in support of the above theoretical predictions (Priestley, 1959; Myrup, 1967; Monin and Yaglom, 1971, 1975; Wyngaard, 1973; Bradley and Antonia, 1979; Panofsky and Dutton, 1984; Stull, 1988; Kaimal and Finnegan, 1994). The above relations may not be valid very close to the surface, because the surface friction and shear-generated turbulence have been ignored in the similarity scaling. They are also not likely to be applicable within the mixed layer, because the PBL height has been ignored. The validity of the local free convection similarity scaling is, thus, restricted to the upper part of the unstable surface layer in the height range $-L < z < 0.1\,h$. For smaller heights or $-z/L$, the Monin-Obukhov similarity theory is found to be valid, except for the horizontal velocity fluctuations. The question arises if there is a smooth transition between the predictions of the two theories as - z/L increases. The answer, according to all the available observational evidence, is yes, but the point of transition on the z/L axis appears to be different for different quantities. This can be examined by expressing Eqs. (26) - (28) in terms of the M-O similarity functions in the local free convection regime:

$$\frac{kz}{\theta_*} \frac{\partial \overline{\theta}}{\partial z} = Ck^{4/3} \left(-\zeta \right)^{-1/3} \tag{29}$$

$$\sigma_w / u_* = C_w k^{-1/3} \left(-\zeta \right)^{1/3} \tag{30}$$

$$\sigma_\theta / \theta_* = C_\theta k^{1/3} \left(-\zeta \right)^{-1/3} \tag{31}$$

Note that the right hand side of Eq. (29) is the predicted form for ϕ_h (ζ) under free convection conditions. A similar expression has been suggested for ϕ_m, although the applicability of free convection scaling to mean wind shear is highly questionable. Equation (29) or its integral form has been proposed and widely used by Russian scientists (Obukhov, 1960; Kazanski and Monin, 1958; Gurvich, 1965; Zilitinkevich and Chalikov, 1968; Monin and Yaglom, 1971). But other investigators did not find compelling evidence in support of the same even for $-\zeta \gg 1$, but limited to 10 (Dyer, 1967; Businger et al., 1971; Businger, 1973; Hogstrom, 1988). The transition to the $\phi_h \sim \left(-\zeta \right)^{-1/3}$ behavior probably occurs for $-\zeta > 10$. But, such large values of - z/L may be attained only at large heights where the potential temperature gradient tends to vanish (Webb, 1958).

The experimental evidence for the validity of Eqs. (30) and (31) is much stronger and the local free convection regime appears to commence when $-\zeta > 0.5$ (Wyngaard et al., 1971; Businger, 1973; Bradley and Antonia, 1979; Kaimal et al., 1982). This is clearly shown by the data in Fig. 3 taken over a dry lake bed (Utah Salt Flats).

3.3 Mixed-layer similarity theory and scaling

In the unstable and convective PBLs, there is a mixed layer above the surface layer $(z \geq 0.1\,h)$ in which the mean potential temperature is observed to be more or less uniform (Kaimal et al., 1976; McBean, 1979; Stull, 1988; Garratt, 1992; Kaimal and Finnegan, 1994). Variations in specific humidity, other scalar concentrations, and mean velocity are also

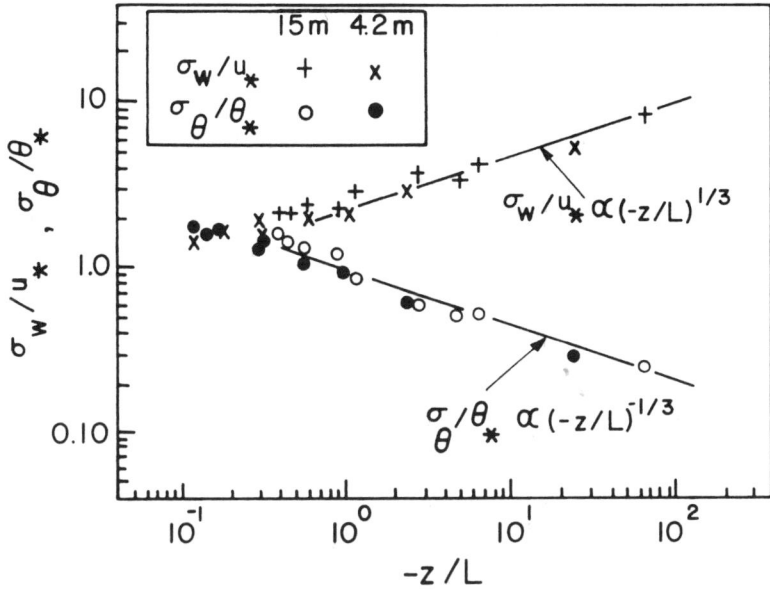

Figure 3: Normalized standard deviations of vertical velocity and temperature fluctuations
           as functions of z/L in the unstable surface layer (from Businger, 1973).

observed to be small over the depth of the mixed layer. Deardorff (1970)
proposed a similarity hypothesis that turbulence structure in the mixed
layer depends only on the surface heat flux $H_o / \rho c_p$, the buoyancy variable
g/T_0, the height above the surface z, and the height of inversion base z_i. The
appropriate scaling parameters for the convective mixed layer are:

length: z_i

velocity: $$W_* = \left(\frac{g}{T_o} \frac{H_o}{\rho c_p} z_i \right)^{1/3}$$ (32)

temperature: $T_* = H_o / \rho c_p W_*$

Here W_* is called the convective velocity and T_* the convective temperature scale. In the definition of W_*, g H_0/T_0 should be replaced by g H_V/T_V, whenever there is a substantial water vapor flux contributing to buoyancy flux.

The mixed-layer similarity theory predicts that the normalized turbulent quantities σ_u / W_*, σ_w / W_*, σ_θ / T_*, etc., must be unique functions of the normalized height z/z_i only. This similarity scaling is strictly valid only within the mixed layer $(0.1 z_i \leq z \leq z_i)$ of the CBL with $-z_i/L > 10$. Observations of turbulence in the CBL (see Fig. 4) indicate, however, that there is hardly any dependence of σ_u / W_* and σ_v / W_* on z / z_i and, for all practical purposes,

$$\sigma_u / W_* \simeq \sigma_v / W_* \simeq 0.60 \qquad (33)$$

For the horizontal velocity components, the mixed-layer scaling is found to be applicable also in the free convective surface layer, where both the Monin-Obukhov similarity and the local free convection similarity theories are found to be wanting. Based on turbulence measurements in unstable surface layers over land and sea, Panofsky et al. (1977) proposed the following interpolation formula for unstable and convective conditions:

$$\sigma_u / u_* \simeq \sigma_v / u_* = (12 - 0.5 z_i / L)^{1/3} \qquad (34)$$

Equation (34) recognizes the observed fact that, under unstable conditions, σ_u and σ_v are independent of height above the surface, but depend on the mixed-layer height. For free convective conditions corresponding to very large values of $-z_i/L$, Eq. (34) is quite consistent with Eq. (33).

The variance of vertical velocity fluctuations in the CBL has been observed to show stronger dependence on height or z/z_i (see Fig. 4). It increases with height, reaches a broad maximum in the middle of the mixed layer, and decreases again in the upper part of the mixed layer. The observed increase in the surface layer is consistent with the local free convection similarity scaling which implies $\overline{w'^2} / W_*^2 \propto (z / z_i)^{2/3}$.

Other statistics of turbulence, such as third moments, probability density functions and wave-number or frequency spectra, are also found to follow the mixed-layer similarity scaling quite well (Kaimal et al., 1976; Stull, 1988; Sorbjan, 1989, 1991; Kaimal and Finnegan, 1994). The wave length corresponding to the peak in spectrum is a measure of the large-eddy

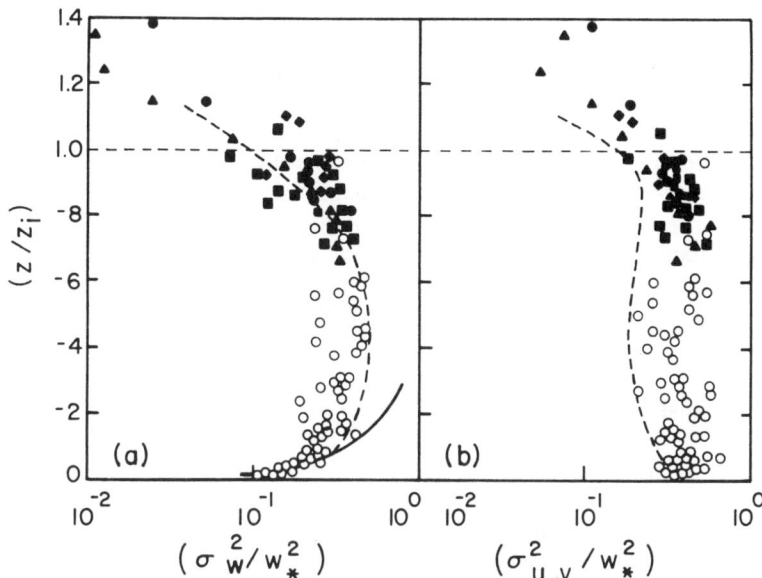

Figure 4:   Variances of horizontal and vertical velocity fluctuations, normalized by W_*^2, as functions of normalized height (z/z_i) in the CBL (from Caughey and Palmer, 1979). The solid line represents the local free convection prediction and dashed lines represent convection tank results of Willis and Deardorff (1974).

length scale. Measurements of the velocity spectra indicate that the peak wavelengths in the horizontal velocity spectra are about 1.5 z_i, independent of height, while the peak wavelength of the vertical spectrum increases with height approaching a value of 1.5 z_i in the upper-half of the mixed layer. The rate of dissipation of the turbulence kinetic energy (TKE) is also found to be approximately constant in the mixed layer. Since the rate of production of TKE decreases linearly with height in proportion to the buoyancy flux, there is a significant vertical transport of energy from the lower-half to the upper-half of the CBL (see e.g., Lenschow, 1974).

Higher moments of turbulence in the mixed layer are also found to be scaled by W_*, although there is considerable amount of scatter in data when plotted in the framework of the mixed-layer similarity theory (Kaimal *et al.*, 1976; Lenschow *et al.*, 1980; Sorbjan, 1989, 1991). In particular, the third moment of vertical velocity fluctuations $\overline{w'^3}$ in the CBL is observed to be positive, indicating larger positive velocities in convective updrafts as compared to negative vertical velocities associated with downdrafts. Since the average vertical velocity is essentially zero in a horizontally-homogeneous boundary layer, the horizontal area occupied by downdrafts is much (about 50%) larger than that occupied by updrafts. This can also be seen from the observed and computed probability density functions of vertical velocity in the CBL (Lamb, 1982; Caughey *et al.*, 1983; Quintarelli, 1990).

3.4 Transition or entrainment layer

Above the mixed layer lies a transition layer in which smooth transition occurs between the convective mixed layer with vigorous turbulence and the free atmosphere with little or no turbulence. During the growing phase of the CBL in morning and early afternoon periods, warmer and drier air from the stable free atmosphere is entrained into the CBL. For this reason, the transition layer is also referred to as the entrainment layer or zone (Deardorff *et al.*, 1980; Stull, 1988). But the former term is more general and applies to non-growing and, hence, non-entraining convective boundary layers as well.

As shown in Fig. 1, the lower part of the transition layer lying below the inversion base is characterized by small positive gradients of $\overline{\theta}$ or $\overline{\theta}_v$, while the upper part has much larger gradients, exceeding even the free-atmospheric value Γ. The buoyancy flux due to entrainment is downward with its maximum value near the inversion base. The negative buoyancy tends to suppress turbulence with increasing height in the transition layer, but the increased wind shear in the transition layer may enhance it. As a result of these opposing influences of buoyancy and shear, horizontal velocity variances, the rate of energy dissipation, and other turbulence parameters often show secondary maxima occurring in the transition layer (see e.g., Sorbjan, 1989, Ch. 4). Due to strong gradients in $\overline{\theta}$, $\overline{q}$, and other scalar profiles in the transition layer near or just above the inversion base, large variances of scalar fluctuations are also expected and have been observed (Willis and Deardorff, 1974; Caughey and Palmer, 1979; Sorbjan, 1989 Ch. 4).

Detailed descriptions of mixed-layer growth and entrainment have been given, among others, by Deardorff *et al.* (1980), Caughey (1982), and Stull

(1988). The average rate of entrainment or the entrainment velocity, w_e, is directly related to the rate of growth of the CBL height as

$$\frac{\partial h}{\partial t} = w_e + \overline{w}_h \qquad (35)$$

where $\overline{w}_h$ is the mean vertical velocity at the top of the CBL. The thickness of the entrainment zone or the transition layer is defined as

$$\Delta h = h - h_o \qquad (36)$$

where h_0 is the particular height near the top of the mixed layer where the buoyancy flux first reaches zero (it becomes zero again at $z = h$).

Based on their laboratory experiments in a convection tank, Deardorff *et al.* (1980) find that the ratios $\Delta h / h_o$ and w_e / W_* are functions of an 'overall' convective Richardson number for the transition layer

$$Ri^* = \frac{g}{\overline{\theta}_v} \frac{\Delta \overline{\theta}_v z_i}{W_*^2} \qquad (37)$$

where $\Delta \overline{\theta}_v$ is the change in the virtual potential temperature across the transition layer or the entrainment zone. Deardorff *et al.* (1980) suggested approximately inverse relationships of the form $\Delta h / h_o \propto (Ri^*)^{-1}$ and $w_e / W_* \propto (Ri^*)^{-1}$, although large scatter in their data does not rule out somewhat different exponent values reported by other investigators. For example, the lidar observations of the entrainment zone in the CBL by Boers and Eloranta (1986) and Boers (1989) suggest a relationship of the form $\Delta h / z_i \propto (Ri^*)^{-1/2}$. Boers points out that the capping inversion was very strong for most of the laboratory tank data while the lidar data was obtained under weaker capping inversions. He suggests that the inversion strength is an additional parameter that probably influences the functional relationship between $\Delta h / z_i$ and Ri^*. Nelson *et al.* (1989) have further shown that any diagnostic relationship between $\Delta h / z_i$ and Ri^* or w_e / W_* cannot be expected to be unique and that one must consider the time dependence of $\Delta h / z_i$, w_e / W_*, and other parameters. Therefore, they proposed a prognostic model for the entrainment layer thickness based on simple thermodynamical considerations.

Nelson *et al.* (1989) show the observed evolution of the various significant heights in the CBL and the entrainment layer thickness for several fair-weather days. They define h_o, z_i and h, respectively, the particular height levels where 10%, 50%, and 100% of air could be identified as the much cleaner, nonturbulent free-atmospheric air which entrained into the CBL. The time evolution of $\Delta h / z_i$ shows a rapid-rise phase in the morning with a large peak (maximum $\Delta h / z_i \sim 0.8 - 1.5$) occurring between 0900 and 1100 local time. Subsequently, $\Delta h / z_i$ decreases and reaches an approximately constant value (0.2 - 0.3) in the afternoon (Nelson *et al.*, 1989). The dimensionless entrainment velocity shows a similarly strong diurnal behavior, but with its peak value occurring about an hour later. Plots of $\Delta h / z_i$ vs. w_e / W_* show a looping or cyclic pattern. These clearly indicate that a diagnostic relationship between $\Delta h / z_i$ and w_e / W_* (alternatively, between $\Delta h / h_o$ and w_e / W_* or Ri^*) is not appropriate and that the time-dependence of the various parameters cannot be ignored. A comparison of the laboratory convection tank data with the atmospheric data on the evolving entrainment layer also suggests that a unique relationship between w_e / W_* and Ri^* may not exist. This raises the question if an appropriate similarity framework can be used for representing the height and structure of the transition layer at the top of an evolving CBL.

In the quest of an appropriate similarity framework for the transition layer, the entrainment velocity (w_e) has been suggested as an important velocity scale, in addition to the convective velocity W_*. Their ratio, w_e / W_*, forms an important similarity parameter. An alternative parameter proposed by Deardorff *et al.* (1980) is Ri^*, defined in Eq. (37), which may not be uniquely related to w_e / W_*, except under certain restricted conditions. Moeng and Wyngaard (1984) used the ratio, $R = \overline{\left(w' \theta'_v\right)}_i / \overline{\left(w' \theta'_v\right)}_o$, of the buoyancy fluxes at $z = z_i$ and $z = 0$ as a surrogate for w_e / W_*. They used the large eddy simulations (LES) of "bottom-up and top-down" diffusion of scalars in the CBL from the infinite area sources at the surface and the top of the CBL, respectively, and expressed the various similarity functions describing the vertical distributions of scalar means, variances and covariances (see also Wyngaard and Brost, 1984).

Sorbjan (1989, 1990, 1991) further extended and utilized the "bottom-up and top-down" decomposition approach in proposing an elaborate local similarity scaling for the CBL, including the transition layer. His proposed similarity scaling is based on two hypotheses: (1) any statistical moment M involving vertical velocity, potential temperature and/or concentration of a passive scalar can be decomposed into two components, i.e., $M = M_b + M_t$, in which M_b denotes the "bottom-up" component which would be equal to M during nonpenetrative convection and M_t represents the "top-down" term which is due to entrainment; (2) each of the two components can be scaled by two sets of local (height-dependent) scales of length, velocity, temperature, etc. The dimensional analysis is used to predict each of the decomposed components in terms of the local scales. The local scales are expressed in terms of the mixed-layer scales and the dimensionless height, assuming linear variations of turbulent fluxes with height. This approach leads to explicit, analytical expressions for the various turbulence moments normalized by mixed-layer scales, each involving two similarity constants and several similarity parameters, viz., z/z_i,

$$R = \left(\overline{w'\,\theta'}\right)_i \Big/ \left(\overline{w'\,\theta'}\right)_o, \ R_q = \left(\overline{w'\,q'}\right)_i \Big/ \left(\overline{w'\,q'}\right)_o, \ \text{and } D = \Delta h \, / \, z_i.$$

Evaluation of the various similarity functions is not easy, however, because the large scatter of experimental data makes it difficult to estimate the various similarity parameters for individual experiments or runs (Sorbjan, 1991). The data from several field and laboratory experiments indicated the following ranges of parameter values:

$$R = 0 \text{ to } -0.7; \ \ R_q = 0 \text{ to } 1.5; \ \ D = 0.1 \text{ to } 1$$

However, Sorbjan (1991) chose the particular values of $R = -0.2$, $R_q = 0.25$ and 0.50, and $D = 0.1$ based on the results of several LES models. The constants in the various similarity relations are estimated from the best fit of theoretical relations to the experimental data. The observed vertical distributions of second and third moments of vertical velocity, temperature and specific humidity are compared with the local similarity predictions. These show fair agreement, but the validation of theory is confounded by large data scatter and large uncertainties in Sorbjan's (1991) estimates of similarity constants and parameters. His hypothesis of local similarity of turbulence structure in the CBL appears to be rather tenuous and should be tested more directly by showing that the various turbulence quantities when made dimensionless by local similarity scales are constant indeed.

A straightforward extension of the mixed-layer similarity theory to the lower part of the transition layer, in which shear-generated turbulence can be ignored, would be to add the time t, and the entrainment velocity w_e to the list of the usual independent variables for the mixed layer (see e.g., Nelson et al., 1989). This gives $t\,W_* / z_i$ and w_e / W_* as independent similarity parameters, in addition to z/z_i. Thus, the appropriate similarity prediction for the entrainment layer thickness is given by

$$\frac{\Delta h}{z_i} = F\left(\frac{t\, W_*}{z_i}\,,\, \frac{w_e}{W_*}\right) \tag{38}$$

Similarly, the various moments of vertical velocity and scalar fluctuations are expected to be functions of z / z_i, w_e / W_*, and $t\, W_* / z_i$.

   In order to properly represent the turbulence structure in the upper part of the transition layer, which is characterized by stronger potential temperature gradient and significant wind shear, one must add the dimensionless stratification parameter, $\Gamma \Delta h / \Delta \bar{\theta}$, and the gradient Richardson number, Ri, to the above list of similarity parameters. This makes the number of similarity parameter too large to be practical. Moreover, the mixed layer scaling may not be appropriate for the upper part of the transition layer. In view of the above considerations, it is doubtful that a generalized similarity framework would be very useful or even practical for describing the upper part of the transition layer or entrainment zone.

4. Mathematical modeling

For practical applications of transport and diffusion processes in the PBL, the various types of mathematical models of the unstable and convective atmospheric boundary layers have been developed. Some of the models require only analytical or simple numerical solutions to the simplified ensemble-averaged equations of motion, thermodynamic energy, etc., while others employ the biggest and fastest supercomputers available to simulate random eddy diffusion processes in three-dimensional space and time. Here, we briefly describe different modeling approaches available in increasing order of their sophistication and complexity. For more detailed derivations of model equations, their closure, and subsequent analytical or numerical solutions representing model results, the reader should refer to more comprehensive reviews of the PBL modeling (Monin and Yaglom, 1971; Haugen, 1973; McBean, 1979; Nieuwstadt and van Dop, 1982; Stull, 1988; Sorbjan, 1989).

4.1 Integral models

The integral or slab models use ensemble-averaged equations which are integrated over the vertical coordinate direction. The integration yields a set of equations for the mixed layer-averaged wind components $(\overline{u}_m$ and $\overline{v}_m)$, potential temperature $(\overline{\theta}_m)$, specific humidity, and other scalar concentrations (see e.g., Driedonks, 1982; Arya and Byun, 1987; Sorbjan, 1989; Stull, 1988). These equations contain fluxes of momentum, heat, etc., at the surface, as well as at the top of the mixed layer or at the inversion base. Closure assumptions are made to parameterize surface fluxes (e.g., through the bulk drag and heat transfer relations), as well as the entrainment fluxes (e.g., through the usual entrainment relations). Rate equations for the mixed-layer depth or the inversion base height, entrainment-layer thickness, and changes in mean variables across the entrainment layer are also carried in integral models. Depending on the sophistication with which the transition or entrainment layer is represented and the type of closure used, the various zero-order and first-order slab models have been proposed for the horizontally-homogeneous CBL (Driedonks, 1982; Arya and Byun, 1987). Simpler zero-order jump models with appropriate closure assumptions may be adequate for predicting the mixed layer depth and potential temperature, but more sophisticated integral models are required for better simulation of the structure in the transition layer.

One of the simplest thermodynamic models of the mixed layer growth during the period of surface heating is based on the solution to the rate equation

$$\partial z_i^2 / \partial t = 2 (1 - R) H_o / \rho c_p \Gamma \qquad (39)$$

in which $R \approx - 0.2$, $\Gamma = \partial \overline{\theta} / \partial z$ in the free atmosphere just above the mixed layer, and the surface heat flux H_0 must be known or specified as a function of time. Equation (39) may be integrated numerically or analytically (e.g., for a sinusoidal variation of H_0 with time). An alternative graphical method can also be used (Stull, 1988, Ch. 11). Although the purely thermodynamic approach neglects turbulent entrainment in the transition layer, it often explains 80-90% of the observed mixed-layer growth.

The integral modeling approach has also been extended and usefully applied to nonhomogeneous surfaces, e.g., for the simulation of urban boundary layers over varying land use and topography, typical of large urban metropolitan areas (Byun, 1987; Byun and Arya, 1990). A particular advantage of integral modeling of mesoscale atmospheric boundary layers over urban and non-urban areas of complex land use patterns and topography is that one only deals with the layer-averaged variables $\left(\text{e.g., } \overline{\theta}_m, \text{ h, } \overline{u}_m, \text{ etc.}\right)$ and their variations with time and one or two spatial dimensions. Such models are much simpler and computationally less expensive than the full three-dimensional mesoscale models. These are most appropriate for unstable and convective conditions in which mean winds, potential temperatures, and other scalar concentrations are expected to attain nearly uniform distributions in the vertical. One of the principal drawbacks of integral models is that they provide no information on the vertical distributions of mean variables and turbulent quantities (e.g., variances and covariances) in the PBL. More sophisticated models have to be used for this purpose.

4.2 First-order closure models

All the turbulence closure models are based on the ensemble-averaged equations of motion, thermodynamic energy, turbulence variances and covariances, etc. The simplest of these are the first-order closure models in which the equations of mean motion $(\overline{u}_i)$, potential temperature $(\overline{\theta})$, and other scalar concentrations are closed and, then, solved for the appropriate initial and boundary conditions. The simpler set of equations for the horizontally-homogeneous PBL are given in section 2; these can easily be generalized to spatially varying non-homogeneous PBLs by including the mean advection terms and other flux divergence terms. In order to close this set of first-order or first-moment equations, turbulent fluxes (also called second moments) are expressed in terms of the mean variables (first moments).

The simplest closure relations are provided by Boussinesq's gradient-transport hypothesis:

$$\overline{u'w'} = -K_m \frac{\partial \overline{u}}{\partial z} \quad ; \quad \overline{v'w'} = -K_m \frac{\partial \overline{v}}{\partial z}$$
$$\overline{\theta'w'} = -K_h \frac{\partial \overline{\theta}}{\partial z} \quad ; \quad \overline{q'w'} = -K_w \frac{\partial \overline{q}}{\partial z} \tag{40}$$

in which K_m, K_h and K_w are the eddy diffusivities or exchange coefficients of momentum, heat and water vapor, respectively. The closure is completed only after K_m, K_h, etc., are specified as constant values or functions of z, u_* and other PBL scaling parameters. Appropriate similarity hypothesis and scaling may be used for this purpose. For example, in the horizontally-homogeneous atmospheric surface layer, the Monin-Obukhov similarity theory suggests

$$K_m = kzu_* / \phi_m (\zeta)$$
$$K_h = kzu_* / \phi_h (\zeta) \tag{41}$$
$$K_w = kzu_* / \phi_w (\zeta)$$

which are often used for modeling transport and diffusion in the surface layer.

A further extension of the gradient-transport (K) theory was provided by Prandtl's mixing length hypothesis, which can be used to derive several alternative expression for eddy viscosity (see e.g., Sutton, 1953; Monin and Yaglom, 1971; Arya, 1988):

$$K_m = \ell_m^2 \left| \frac{\partial \overline{V}}{\partial z} \right| \tag{42}$$

$$K_m = c_m \ell_m \sigma_w \tag{43}$$

where $\left| \overline{V} \right|$ is the mean wind vector, ℓ_m is the mean mixing length for momentum, and c_m is an empirical constant of the order one. Similar expressions can be written for K_h and K_w by introducing mean mixing lengths for heat and water vapor. Then, for complete closure, the various mixing lengths have to be specified. The most commonly used expression for the mixing length in the PBL is

$$\ell_m = kz \left(1 + kz / \ell_\infty \right)^{-1} \tag{44}$$

which gives an interpolation between its expected linear increase with height near the surface and the constant value ℓ_∞ in the upper part of the PBL. The latter may be specified in terms of the PBL height (e.g., $\ell_\infty \propto h$) or the vertical distribution of turbulence kinetic energy (Mellor and Yamada, 1974). The latter can be obtained from the TKE equation which is really a second-moment equation.

The use of the turbulence kinetic energy (TKE) equation in otherwise first-order closure models is also frequently made for specifying the eddy viscosity:

$$K_m = c \, \ell \, E^{1/2} \qquad (45)$$

where $E = \left(\overline{u'^2} + \overline{v'^2} + \overline{w'^2} \right) / 2$ is the TKE, ℓ is the large-eddy length scale which is proportional to the mixing length and can be specified in the same manner, and c is an empirical constant. The rate of energy dissipation (ε) in the TKE or E-equation is usually parameterized as

$$\varepsilon = c_\varepsilon \, E^{3/2} / \ell_\varepsilon \qquad (46)$$

where ℓ_ε is the dissipation length scale which is similar to but may differ in magnitude from ℓ, and c_ε is another empirical constant which is related to c (equating the energy dissipation rate to the shear production in the neutral surface layer gives $c_\varepsilon = c^3$). Alternatively, a much simplified and heavily parameterized form of the equation for ε is included in the so-called E-ε or k-ε models (Detering and Etling, 1985; Holt and Raman, 1988; Huang and Raman, 1989). Consequently, several additional empirical constants have also to be specified; these have been evaluated largely from the laboratory experimental data and may not be appropriate for the unstable and convective atmospheric boundary layers. A comprehensive review of the first-order closure models, including the TKE and E-ε models (these are sometimes referred to as 1-1/2 closure models) is given by Holt and Raman (1988).

Although the gradient-transport (K) type first-order closure models, utilizing a variety of eddy diffusivity and mixing length parameterizations, have been extensively used for practical applications in micrometeorology and air pollution dispersion, their validity under very unstable and convective conditions has been questioned. In particular, turbulent transports or fluxes are not found to be related to the vanishing small local mean gradients in the convective mixed layer. For example, in the middle and upper parts of the mixed layer, $\partial \overline{\theta} / \partial z$ is near zero or slightly positive while the turbulent heat flux is upward implying infinite or negative values of K_h. Similar counter-gradient fluxes of other scalars, as well as momentum, have also been observed in unstable and convective boundary layers. The local gradient-transport (K and ℓ) hypotheses become generally invalid as they are found to be inconsistent with the observed turbulence structure.

Considering the simplicity and versatility of the first-order closure approach, several nonlocal closure methods have been proposed for modeling turbulence and diffusion in the CBL (Stull, 1984, 1988, 1993; Brown, 1993; Brown and Arya, 1995). Nonlocal closures are based on the concept that all turbulent eddies can transport flow properties across finite distances through eddies of varying sizes. Consequently, mean velocity, potential temperature and other flow properties at any location are a result of transport and mixing of fluid parcels coming from many cross-stream (along the gradients of mean properties) locations.

The basis of nonlocal closure models can be explained by considering turbulent mixing of parcels carried by eddies of different sizes in the vertical direction only (Stull, 1988; Sorbjan, 1989). The vertical column of air may be assumed to be divided between finite number N of equal size grid boxes. Turbulent mixing of fluid coming from any box (denoted by index j) into the reference grid box (denoted by index i) can change any conserved passive scalar property s_i (potential temperature, humidity, etc.) of the reference box. If c_{ij} represents the fraction of air in box i that came from box j during a time interval Δt, one can write

$$\bar{s}_i\,(t + \Delta t) = \sum_{j=1}^{N} c_{ij}\,(t,\,\Delta t)\,\bar{s}_j\,(t) \qquad (47)$$

where the mixing coefficients c_{ij} describe mixing between all possible pairs of grid boxes. By definition, the conservation of air mass and tracer amount requires that

$$\sum_{j=1}^{N} c_{ij} = \sum_{i=1}^{N} c_{ij} = 1 \qquad (48)$$

The vertical turbulent flux of scalar across any height level k (defined as the top of the grid box k) is given by (Stull, 1988)

$$\overline{w's'}\,(k) = \frac{\Delta z}{\Delta t} \sum_{i=1}^{k} \sum_{j=1}^{N} c_{ij}\left(\bar{s}_i - \bar{s}_j\right) \qquad (49)$$

where Δz is the grid spacing and Δt is the time step interval for the c_{ij} $(t, \Delta t)$ matrix.

The above discrete-grid expression for the scalar flux can also be written in a continuous integral form (Stull, 1988):

$$\overline{w's'}\,(z,\,t) = \int_{z=z_b}^{z_t} \int_{Z=z_b}^{z_t} \gamma\,(z,\,Z,\,t)\,\left[\bar{s}\,(Z,\,t)\,-\,\bar{s}\,(z,\,t)\right]dZ\,dz \qquad (50)$$

where $\gamma\,(z,\,Z,\,t)$ is a rate coefficient of dimensions $\left[L^{-1}\,T^{-1}\right]$ for mixing between levels Z and z and the heights z_b and z_t represent the bottom and top of the turbulent domain. By definition, turbulent fluxes at the top and bottom of the turbulent domain are zero, but the surface flux can be expressed entirely as a molecular flux. The principal drawback of Stull's transilient turbulence model is that the whole matrix of mixing coefficients or mixing rate coefficients has to be specified. These are expected to be different for different flow situations and cannot be generalized. Different approaches have been suggested for parameterizing the mixing coefficients c_{ij}, but none of them is very rigorous and completely satisfactory (Stull, 1988, 1993).

Other types of nonlocal first-order closure models proposed in the literature are the spectral eddy diffusivity model in which K varies with eddy size or wave number (Bercowicz and Prahm, 1980; Bercowicz, 1984), and the nonlocal mixing model of Brown (1993) for computing the variances and probability density functions of velocity and scalar fluctuations (Brown and Arya, 1995).

4.3 Higher-order closure models

Knowing the serious limitations and conceptual flaws of the local gradient-transport theory, when applied to unstable and convective boundary layers, a variety of second-order closure models have been formulated and used for modeling turbulence and diffusion in the CBL (see e.g., Wyngaard and Coté, 1974; Wyngaard et al., 1974a, 1974b; Mellor and Yamada, 1974, 1982; Sun and Chang, 1986). Second-order closure models carry the dynamical equations for the second moments (variances and fluxes), in addition to those of the first moments. This set of equations contain the unknown third moments, correlations involving pressure and velocity or other scalar fluctuations, and the molecular dissipation terms, which have to be parameterized in order to close them. The derivation of second-moment equations and some of the fundamental closure assumptions go back to the 1940s and early 1950s (for a more comprehensive historical review, see Monin and Yaglom, 1971). But, numerical solutions of these equations for applications to PBL modeling had to wait for the later developments in large digital computers and computational fluid dynamics.

Different closure approaches have been proposed for modeling (parameterizing) turbulent transport, pressure, and dissipation terms that appear in second-moment equations (Donaldson, 1973; Mellor and Yamada, 1974, 1982; Lumley and Khajeh-Nouri, 1974; Lumley, 1979; Wyngaard, 1982). Mellor and Yamada (1974) developed a hierarchy of second-order closure models, depending on the type of approximations made in the second-moment equations and complexity of closure assumptions. Their most complete level-4 model contains full set of time-dependent partial differential equations, while the simplest level-1 model is equivalent to a gradient-transport (K) model. Most of the closure assumptions used in different second-order models are quite similar and differ only in their specifications of the various empirical constants and length or time scales.

An accurate representation or parameterization of the molecular dissipation terms in second-moment equations is important because, apart from setting the rate of energy loss to small dissipating eddies, ε determines the turbulence time and length scales. In the simplest approach, ε, ε_θ, etc., are expressed in terms of the appropriate turbulence velocity and length scales as in Eq. (46). The former is equal or proportional to the turbulence kinetic energy which is computed in the model, while the length scale is specified explicitly or through a dynamical equation (Mellor and Yamada, 1982). Alternatively, dynamical equations are also carried for the rates of dissipation of TKE and variances of temperature and other scales. Such equations are very complex (Lumley, 1979) and only greatly simplified and highly parameterized forms of the equations are used (Wyngaard et al., 1974a, 1974b; Lumley, 1979). The molecular dissipation terms in the dynamical equations for the covariances or fluxes are usually neglected, using Kolmogorov's local-isotropy hypothesis for large Reynolds-number flows. The same hypothesis also implies $\varepsilon_u = \varepsilon_v = \varepsilon_w = \varepsilon/3$.

The satisfactory parameterization of the covariances between fluctuating pressure and fluctuating velocity or temperature gradients, which are also called return-to-isotropy terms, is probably the most difficult problem of higher-order closure modeling. The fluctuating pressure field in a stratified shear flow, such as the PBL, can be expressed entirely in terms of the fluctuating velocity and temperature fields and mean velocity shear. The resulting Poisson equation for p' clearly indicates that pressure fluctuations at any point depend on the velocity and temperature fields in a large neighborhood around that point (Lumley, 1979; Zeman, 1981; Wyngaard, 1982). Formally, the return-to-isotropy terms constitute three distinct contributions due to   (1) nonlinear turbulence-turbulence interactions, (2) mean shear-turbulence interactions, and   (3) buoyancy-turbulence interactions. One can probably ignore the relatively minor contribution of the Coriolis or rotational terms.

In the simplest parameterizations (Mellor and Yamada, 1974; Wyngaard *et al.*, 1974a,b), pressure terms are assumed to be proportional to the second moments they tend to destroy and return to isotropy. But this approach, originally proposed by Rotta (1951), is found to be suitable only for the turbulence-turbulence interaction part. More complex parameterizations have been obtained for the contributions due to mean shear and buoyancy (Lumley, 1979; Zeman, 1981). Needless to say, more complex parameterizations based on Lumley's functional expansion technique also involve more empirical constants some of which can be determined by applying symmetry, incompressibility, and integral constraints. Unfortunately, these are not always found to be consistent with the observed mean flow and turbulence structure in the PBL (Zeman, 1981). A satisfactory, but simple parameterization of pressure terms in second-moment equations is still not available.

Turbulent transport terms are the third-moment terms appearing in second-moment equations, which cause the basic closure problem of the latter.   While these are not found very significant in neutral and stably stratified PBLs, the vertical turbulent transports are extremely important and often govern the flow dynamics in unstable and convective boundary layers. Their accurate parameterization is crucial to the proper modeling of the CBL using the second-order closure approach.

The simplest parameterization of turbulent transport is based on the gradient diffusion concept, whereby third moments are expressed in terms of the products of the gradients of second moments and some form of turbulent diffusivity which is parameterized in terms of length and velocity scales (Donaldson, 1973; Mellor and Yamada, 1974; Wyngaard *et al.*, 1974b). However, this type of turbulent-transport modeling has been found to be deficient and inconsistent with the observed turbulence structure in the CBL.   Buoyancy-driven turbulence is capable of maintaining counter-gradient transport.

A more complex functional-expansion approach (Lumley and Khajeh-Nouri, 1974) can remove most of the deficiencies of the simpler approach, but introduces many more terms and empirical constants whose values cannot be determined from the available experimental data.   Another approach is to carry the dynamical equations for third moments, but these have their own closure problems (Zeman and Lumley, 1976; André *et al.*, 1978; Lumley, 1979; Zeman, 1981).

4.4 Large-eddy simulation

Except for the direct numerical simulation (DNS) which aims at resolving even the smallest of eddies and is currently amenable to only very low Reynolds-number turbulent flows, three-dimensional large eddy simulation (LES) is considered to be the best and most fundamental approach to turbulence modeling. Its application to PBL modeling was pioneered by Deardorff (1970, 1972, 1973, 1974). LES aims to resolve explicitly large-scale turbulent eddies that carry most of the turbulence kinetic energy and fluxes of momentum, heat, etc., and to parameterize the effects of small 'subgrid' scale (SGS) motions. Thus, the LES approach occupies the middle ground between the direct numerical simulation and ensemble-averaged closure modeling. It would approach the former ideal as computers become more and more powerful allowing for increasing number of grid points and reduced grid spacing, thus permitting explicit resolution of smaller and smaller eddies. With the currently available supercomputers with parallel-processing architecture, eddy sizes down to a few meters can be resolved over limited domains of a few hundred meters.

The LES approach has been particularly useful for modeling unstable and convective boundary layers, because unresolved SGS motions make only small contributions to transport and diffusion processes and even crude parameterizations of the same may be adequate. Consequently, the simplest non-linear eddy viscosity model proposed by Smagorinsky (1963) works as well as higher-order SGS-closure models (Nieuwstadt et al., 1991). Considering the expenditure of additional computational resources due to increased sophistication of SGS modeling, increased resolution with simpler parameterization might be preferred (Nieuwstadt et al., 1992; Mason, 1994).

The set of equations of motion, potential temperature, etc., used in LES is obtained by grid volume-averaging of the instantaneous equations. This amounts to applying a spatial filter, which for all practical purposes is fixed by the numerical grid resolution. Several SGS models have been proposed with varying levels of sophistication (Deardorff, 1973; Moeng, 1984; Schmidt and Schumann, 1989; Mason, 1989; Moeng and Sullivan, 1994).

The well-known Smagorinsky model expresses the subgrid scale momentum flux as

$$\langle u_i \, u_j \rangle - \langle u_i \rangle \langle u_j \rangle = K'_m \left(\frac{\partial \langle u_i \rangle}{\partial x_j} + \frac{\partial \langle u_j \rangle}{\partial x_i} \right) \tag{51}$$

where K'_m is the subgrid scale eddy viscosity which is given by

$$K'_m = \ell_o^2 S_{ij} \tag{52}$$

and

$$S_{ij}^2 = \frac{1}{2} \left(\frac{\partial \langle u_i \rangle}{\partial x_j} + \frac{\partial \langle u_j \rangle}{\partial x_i} \right)^2 \tag{53}$$

Here $< >$ denotes the grid-volume averaging filter operation and ℓ_o is a mixing-length scale which is related to the filter operation and, hence, to the average grid size as

$$\ell_o = C_s \Delta = C_s \left(\Delta x\, \Delta y\, \Delta z \right)^{1/3} \tag{54}$$

C_S is a key parameter of the above SGS closure. Although a value of $C_s \simeq 0.17$ was derived by Lilly (1967) using the inertial subrange relations for SGS motions, it has actually been found to be dependent on the filter scale (Mason, 1994).

A long-standing problem in LES has been the inadequate resolution of the gradients in the lower part of the surface layer. Consequently, the well known logarithmic velocity profile in the neutral surface layer and the corresponding Monin-Obukhov similarity relations for stratified surface layer are not recovered. This situation has been rectified, to some extent, by modifying the SGS models close to the surface where even large energy containing eddies become subgrid scale (see e.g., Sullivan *et al.*, 1994).

References

André, J.C., G. DeMoor, P. Lacarrere, G. Therry and R. Du Vachat, 1978: Modeling the 24-hour evolution of the mean and turbulent structures of the planetary boundary layer. *J. Atmos. Sci.*, *35*, 1861-1883.

Arya, S.P., 1988: *Introduction to Micrometeorology*, Academic Press, San Diego, 307 pp.

Arya, S.P. and D.W. Byun, 1987: Rate equations for the planetary boundary layer depth (urban vs rural). *Modeling the Urban Boundary Layer*, American Meteorological Society, Boston, 215-252.

Bercowicz, R., 1984: Spectral methods for atmospheric diffusion. *Boundary-Layer Meteor.*, *30*, 201-220.

Bercowicz, R. and L.P. Prahm, 1980: On spectral turbulent theory for homogeneous turbulence. *J. Fluid Mech.*, *100*, 433-448.

Boers, R., 1989: A parameterization of the depth of the entrainment zone. *J. Appl. Meteor.*, *28*, 107-111.

Boers, R. and E.W. Eloranta, 1986: Lidar measurements of the atmospheric entrainment zone and the potential temperature jump across the top of the mixed layer. *Boundary-Layer Meteor.*, *34*, 357-375.

Bradley, E.F. and R.A. Antonia, 1979: Structure parameters in the atmospheric surface layer. *Quart. J. Roy. Meteor. Soc.*, *105*, 695-705.

Brown, M.J., 1993: A nonlocal model for prediction of the probability density of turbulent fluctuations in boundary-layer flows. Ph.D. Dissertation, North Carolina State University, Raleigh, NC, 143 pp.

Brown, M.J. and S.P. Arya, 1995: Introduction to a non-local mixing model for prediction of the probability density function of velocity fluctuations in boundary-layer flows. Preprints, 11th Symposium on Boundary Layers and Turbulence, American Meteorological Society, Boston, MA, 144-147.

Businger, J.A., 1973: Turbulent transfer in the atmospheric surface layer. *Workshop on Micrometeorology*, D.A. Haugen, ed., American Meteorological Society, Boston, MA, 67-100.

Businger, J.A., J.C. Wyngaard, Y. Izumi, and E.F. Bradley, 1971: Flux-profile relationship in the atmospheric surface layer. *J. Atmos. Sci.*, *28*, 181-189.

Byun, D.W., 1987: A two-dimensional numerical model of St. Louis urban mixed layer, Ph.D. Dissertation, North Carolina State University, Raleigh, NC, 216 pp.

Byun, D.W. and S.P. Arya, 1990: A two-dimensional mesoscale numerical model of an urban mixed layer – I. Model formulation, surface energy budget, and mixed-layer dynamics. *Atmos. Environ.*, *24A*, 829-844.

Caughey, S.J., 1982: Observed characteristics of the atmospheric boundary layer. *Atmospheric Turbulence and Air Pollution Modelling*, F.T.M. Nieuwstadt and H. van Dop, editors, D. Reidel Pub. Co., Dordrecht, Holland, 107-158.

Caughey, S.J. and S.G. Palmer, 1979: Some aspects of turbulence structure through the depth of the convective boundary layer. *Quart. J. Roy. Meteor. Soc.*, *105*, 811-827.

Caughey, S.J., M. Kitchen, and J.R. Leighton, 1983: Turbulence structure in convective boundary layers and implications for diffusion. *Boundary-Layer Meteor.*, *25*, 345-352.

Deardorff, J.W., 1970: Convective velocity and temperature scales for the unstable planetary boundary layer and for Rayleigh convection. *J. Atmos. Sci.*, *27*, 1211-1213.

Deardorff, J.W., 1972: Numerical investigation of neutral and unstable planetary boundary layer. *J. Atmos. Sci.*, *29*, 91-115.

Deardorff, J.W., 1973: Three-dimensional numerical modeling of the planetary boundary layer. *Workshop on Micrometeorology*, D.A. Haugen, editor, American Meteorological Society, Boston, MA, 271-311.

Deardorff, J.W., 1974: Three-dimensional study of turbulence in an entraining mixed layer. *Boundary-Layer Meteor., 7*, 199-226.

Deardorff, J.W., G.E. Willis, and B.H. Stockton, 1980: Laboratory studies of the entrainment zone of a convective mixed layer. *J. Fluid Mech., 100*, 41-64.

Donaldson, C. du P., 1973: Construction of a dynamic model of the production of atmospheric turbulence and the dispersal of atmospheric pollutants. *Workshop on Micrometeorology*, D.A. Haugen, editor, American Meteorological Society, Boston, MA, 313-392.

Detering, H.W. and D. Etling, 1985: Application of the E-ε turbulence model to the atmospheric boundary layer. *Boundary-Layer Meteor., 33*, 113-133.

Driedonks, A.G.M., 1982: Models and observations of the growth of the atmospheric boundary layer. *Boundary-Layer Meteor., 23*, 283-306.

Dyer, A.J., 1967: The turbulent transport of heat and water vapor in unstable atmosphere. *Quart. J. Roy. Meteor. Soc., 93*, 501-508.

Dyer, A.J., 1974: A review of flux-profile relationships. *Boundary-Layer Meteor., 7*, 363-372.

Dyer, A.J. and B.B. Hicks, 1970: Flux-gradient relationships in the constant flux layer. *Quart. J. Roy. Meteor. Soc., 96*, 715-721.

Garratt, J.R., 1992: *The Atmospheric Boundary Layer*, Cambridge University Press, Cambridge, 316 pp.

Garratt, J.R. and B.B. Hicks, 1990: Micrometeorological and PBL experiments in Australia. *Boundary-Layer Meteor., 50*, 11-29.

Gurvich, A.S., 1960: Frequency spectra and functions of distribution of probabilities of vertical wind velocity components. *Izv. Geophys. Ser., No. 7*, 695-703 (English Transl.).

Gurvich, A.S., 1965: Vertical temperature and wind velocity profiles in the atmospheric surface layer. *Izv. Atmos. and Ocean. Phys., 1*, 31-36 (English Transl.).

Haugen, D.A., 1973: *Workshop on Micrometeorology*, American Meteorological Society, Boston, MA, 392 pp.

Hicks, B.B., 1976: Wind profile relationships from the 'Wangara' experiment. *Quart. J. Roy. Meteor. Soc., 102*, 535-551.

Hogstrom, U., 1982: A critical evaluation of the aerodynamic error of a turbulence instrument. *J. Appl. Meteor., 21*, 1838-1844.

Hogstrom, U., 1985: von Karman's constant in atmospheric boundary layer flow: reevaluated. *J. Atmos. Sci., 42*, 263-270.

Hogstrom, U., 1988: Non-dimensional wind and temperature profiles in the atmospheric surface layer: a reevaluation. *Boundary-Layer Meteor., 42*, 55-78.

Holt, T. and S. Raman, 1988: A review and comparative evaluation of multilevel boundary layer parameterizations for first-order and turbulent kinetic energy closure schemes. *Reviews of Geophys., 26*, 761-780.

Huang, C.Y. and S. Raman, 1989: Application of the E-ε closure model to simulations of mesoscale topographic effects. *Boundary-Layer Meteor., 49*, 169-195.

Kaimal, J.C. and J.J. Finnegan, 1994: *Atmospheric Boundary Layer Flows*. Oxford University Press, New York.

Kaimal, J.C., J.C. Wyngaard, Y. Izumi, and O.R. Coté, 1972: Spectral characteristics of surface layer turbulence. *Quart. J. Roy. Meteor. Soc.*, *98*, 563-589.

Kaimal, J.C., J.C. Wyngaard, D.A. Haugen, O.R. Coté, Y. Izumi, S.J. Caughey, and C.J. Readings, 1976: Turbulence structure in the convective boundary layer. *J. Atmos. Sci.*, *33*, 2152-2168.

Kaimal, J.C., R.A. Eversole, D.H. Lenschow, B.B. Stankov, P.H. Kahn, and J.A. Businger, 1982: Spectral characteristics of the convective boundary layer over uneven terrain. *J. Atmos. Sci.*, *39*, 1098-1114.

Kazanski, A.B. and A.S. Monin, 1958: The turbulent regime in the air layer next to the ground with unstable stratification. *Izv. Acad. Sci. USSR, Geophys. Ser., No. 6*, 741-751.

Kondo, J. and T. Sato, 1982: The determination of the von Karman constant. *J. Meteor. Soc. Japan*, *60*, 461-470.

Lamb, R.G., 1982: Diffusion in the convective boundary layer. Atmospheric Turbulence and Air Pollution Modelling, F.T.M. Nieuwstadt and H. van Dop, ed., D. Reidel Pub. Co., Dordrecht, Holland, 159-229.

Lenschow, D.H., 1974: Model of the height variation of the turbulent kinetic energy budget in the unstable planetary boundary layer. *J. Atmos. Sci.*, *31*, 465-474.

Lenschow, D.H., J.C. Wyngaard, and W.T. Pennel, 1980: Mean field and second-moment budgets in a baroclinic convective boundary layer. *J. Atmos. Sci.*, *37*, 1313-1326.

Lilly, D.K., 1967: The representation of small-scale turbulence in numerical simulation experiments. Proc. Tenth IBM Scientific Computing Symposium on Environmental Sciences, Thomas J. Watson Research Center, Yorktown Heights, NY, 195-210.

Lumley, J.L., 1979: Computational modeling of turbulent flows. *Adv. Appl. Mech.*, *18*, 123-176.

Lumley, J.L. and B. Khajeh-Nouri, 1974: Computational modeling of turbulent transport. *Adv. Geophys.*, *18A*, 169-192.

Lumley, J.L. and H.A. Panofsky, 1964: *The Structure of Atmospheric Turbulence*. Interscience Pub., New York, 239 pp.

Mason, P.J., 1989: Large-eddy simulation of the convective atmospheric boundary layer. *J. Atmos. Sci.*, *46*, 1492-1516.

Mason, P.J., 1994: Large-eddy simulation: a critical review of the technique. *Quart. J. Roy. Meteor. Soc.*, *120*, 1-26.

McBean, G.A., 1979: *The Planetary Boundary Layer*. Technical Note No. 165, World Meteorological Organization, Geneva, 201 pp.

Mellor, G.L. and T. Yamada, 1974: A hierarchy of turbulence closure models for planetary boundary layers. *J. Atmos. Sci.*, *31*, 1791-1806.

Mellor, G.L. and T. Yamada, 1982: Development of a turbulence closure model for geophysical fluid problems. *Rev. Geophys. Space Phys.*, *20*, 851-875.

Moeng, C.H., 1984: A large-eddy simulation model for the study of planetary boundary-layer turbulence. *J. Atmos. Sci.*, *41*, 2052-2062.

Moeng, C.H. and P.P. Sullivan, 1994: A comparison of shear and buoyancy-driven planetary boundary layer flows. *J. Atmos. Sci.*, *71*, 999-1022.

Moeng, C.H. and J.C. Wyngaard, 1984: Statistics of conservative scalars in the convective boundary layer. *J. Atmos. Sci.*, *41*, 3161-3169.

Monin, A.S. and A.M. Obukhov, 1954: Basic laws of turbulent mixing in the atmosphere near the ground. *Tr. Akad. Neuk SSSR Geofiz. Inst.*, *No. 24* (151), 163-187.

Monin, A.S. and A.M. Yaglom, 1971: *Statistical Fluid Mechanics*: *Mechanics of Turbulence, Vol. 1*, MIT Press, Cambridge, MA, 769 pp.

Monin, A.S. and A.M. Yaglom, 1975: *Statistical Fluid Mechanics*: *Mechanics of Turbulence, Vol. 2*, MIT Press, Cambridge, MA, 874 pp.

Myrup, L.O., 1967: Temperature and vertical velocity fluctuations in strong convection. *Quart. J. Roy. Meteor. Soc.*, *93*, 350-360.

Nelson, E., R. Stull, and E. Eloranta, 1989: A prognostic relationship for entrainment zone thickness. *J. Appl. Meteor.*, *28*, 885-903.

Nieuwstadt, F.T.M. and H. van Dop, 1982: *Atmospheric Turbulence and Air Pollution Modelling*. D. Reidel Pub. Co., Dordrecht, Holland, 358 pp.

Nieuwstadt, F.T.M., P.J. Mason, C.H. Moeng, and U. Schumann, 1992: Large-eddy simulation of the convective boundary layer: A comparison of four computer codes. *Turbulent Shear Flows*, *8*, F. Durst et al., ed., Springer-Verlag, Berlin, 343-367.

Obukhov, A.M., 1946: Turbulence in an atmosphere with a non-uniform temperature. *Tr. Inst. Teor. Geofiz., Akad. Nauk SSSR*, *1*, 95-115.

Obukhov, A.M., 1960: The structure of the temperature and velocity fields in free convection. *Bull. (Izv.) Acad. Sci. USSR, Geophys. Ser.*, 928-930 (English Transl.).

Oke, T.R., 1987: *Boundary Layer Climates, Second Edition*, Methuen, London, 435 pp.

Panofsky, H.A. and J.A. Dutton, 1984: *Atmospheric Turbulence*. Wiley Interscience, New York, 397 pp.

Panofsky, H.A., H. Tennekes, D.H. Lenschow, and J.C. Wyngaard, 1977: The characteristics of turbulent components in the surface layer under convective conditions. *Boundary-Layer Meteor.*, *11*, 355-361.

Paulson, C.A., E. Leavitt, and R.G. Fleagle, 1972: Air-sea transfer of momentum, heat and water determined from profile measurements during BOMEX. *J. Phys. Oceanogr. 2*, 487-497.

Priestley, C.H.B., 1959: *Turbulent Transfer in the Lower Atmosphere*. The Univ. of Chicago Press, Chicago, 130 pp.

Quintarelli, F., 1990: A study of vertical velocity distributions in the planetary boundary layer. *Boundary-Layer Meteor.*, *52*, 209-219.

Rotta, J.C., 1951: Statistische Theorie nichthamogener Turbulenz. *Z. Phys.*, *129*, 547-572.

Schmidt, H. and U. Schumann, 1989: Coherent structure of the convective boundary layer derived from large-eddy simulations. *J. Fluid Mech.*, *200*, 511-562.

Smagorinsky, J., 1963: General circulation experiments with the primitive equations. I. The basic experiment. *Mon. Weath. Rev.*, *91*, 99-164.

Sorbjan, Z., 1989: *Structure of the Atmospheric Boundary Layer*. Prentice-Hall, Englewood Cliffs, NJ, 317 pp.

Sorbjan, Z., 1990: Similarity scales and universal profiles of statistical moments in the convective boundary layer. *J. Appl. Meteor.*, *29*, 762-775.

Sorbjan, Z., 1991: Evaluation of local similarity functions in the convective boundary layer. *J. Appl. Meteor.*, *30*, 1565-1583.

Stull, R.B., 1984: Transilient turbulence theory. Part 1. The concept of eddy-mixing across finite distances. *J. Atmos. Sci.*, *41*, 3351-3367.

Stull, RB., 1988: *An Introduction to Boundary Layer Meteorology*. Kluwer Acad. Pub., Dordrecht, Holland, 666 pp.

Stull, RB., 1993: Review of transilient turbulence theory and non-local mixing. *Boundary-Layer Meteor.*, *62*, 21-96.

Sullivan, P.P., J.C. McWilliams, and C.H. Moeng, 1994: A subgrid-scale model for large-eddy simulation of planetary boundary layer flows. *Boundary-Layer Meteor.*, *71*, 247-276.

Sun, W.Y. and C.Z. Chang, 1986: Diffusion model for a convective boundary layer. Part 1. Numerical simulation of convective boundary layer. *J. Climate Appl. Meteor.*, *25*, 1445-1453.

Sutton, O.G., 1953: *Micrometeorology*. McGraw-Hill, New York.

Verma, S.B., N.J. Rosenberg, and B.L. Blad, 1978: Turbulent exchange coefficients for sensible heat and water vapor under advective conditions. *J. Appl. Meteor.*, *17*, 330-338.

Webb, E.K., 1958: Vanishing potential temperature gradients in strong convection. *Quart. J. Roy. Meteor. Soc.*, *84*, 118-125.

Wieringa, J., 1980: A reevaluation of the Kansas mast influence on measurements of stress and cup anemometer overspeeding. *Boundary-Layer Meteor.*, *18*, 411-430.

the unstable planetary boundary layer. *J. Atmos. Sci.*, *31*, 1297-1307.

Wyngaard, J.C., 1973: On surface layer turbulence. *Workshop on Micrometeorology*, D.A. Haugen, ed., American Meteorological Society, Boston, 101-148.

Wyngaard, J.C., 1982: Boundary-Layer Modeling. *Atmospheric Turbulence and Air Pollution Modelling*, F.T.M. Nieuwstadt and H. vanDop, ed., D. Reidel Pub. Co., Dordrecht, Holland, 69-106.

Wyngaard, J.C. and R.A. Brost, 1984: Top-down and bottom-up diffusion of a scalar in the convective boundary layer. *J. Atmos. Sci.*, *41*, 102-112.

Wyngaard, J.C. and O.R. Coté, 1974: The evolution of a convective planetary layer – a higher-order-closure study. *Boundary-Layer Meteor.*, *7*, 284-308.

Wyngaard, J.C., S.P.S. Arya, and O.R. Coté, 1974a: Some aspects of the structure of convective planetary boundary layers. *J. Atmos. Sci.*, *31*, 747-754.

Wyngaard, J.C., O.R. Coté, and K.S Rao, 1974b: Modeling the atmospheric boundary layer. *Adv. Geophys.*, *18A*, 193-211.

Yaglom, A.M., 1977: Comments on wind- and temperature flux-profile relationships. *Boundary-Layer Meteor.*, *11*, 89-102.

Zeman, O., 1981: Progress in modeling of planetary boundary layers. *Ann. Rev. Fluid Mech.*, *13*, 253-272.

Zeman, O. and J.L. Lumley, 1976: Modeling buoyancy-driven mixed layers. *J. Atmos. Sci.*, *33*, 1974-1988.

Zilitinkevich, S.S. and D.V. Chalikov, 1968: Determining the universal wind-velocity and temperature profiles in the atmospheric boundary layer. *Izv. Atmos. Ocean. Phys.*, *4*, 165-169 (English Transl.).

Chapter 2
Turbulence and dispersion in the stable atmospheric boundary layer
Section A

K. Shankar Rao & Carmen J Nappo

Atmospheric Turbulence and Diffusion Division
Air Resources Laboratory, NOAA
Oak Ridge, TN 37831, USA

Abstract

The stable atmospheric boundary layer over land at night is much shallower, and the turbulent transfers in it are much weaker than in the daytime convective boundary layer. The stable boundary layer is inherently more complex, and our understanding of its structure is limited by difficulties in measurements and modeling. Yet, stable boundary layers are very important in several disciplines of fluid mechanics, including atmospheric dispersion. In this chapter, the turbulent structure of the idealized stable atmospheric boundary layer is described in terms of similarity theories and observations from field experiments. The traditional and new approaches to modeling of turbulence and dispersion in the stable boundary layer are outlined. Several complicating factors, such as radiative transfer, gravity waves, intermittency, and measurement problems are discussed.

1 Introduction

In this chapter, we provide a description of selected features of the stably-stratified atmospheric boundary layer (ABL) over land at night. Details of the idealized stable boundary layer (SBL) can be found, for example, in Arya (1988), Stull (1988), Sorbjan (1989), and Garratt (1992). In our conceptual model, the ground surface is flat, homogeneous, and has a surface roughness z_o. The atmospheric flow is cloud-free, horizontally uniform, barotropic, and is characterized by density $\rho(z)$ and a geostrophic wind G directed along the x-axis; however, the wind vector at the ground surface is oriented at an angle α to the x-axis. The model is best applied at middle latitudes where the Coriolis force, $\rho f G$, is effective. Here, f is the Coriolis parameter.

Shortly before sunset, radiational heating at the ground surface becomes less than the loss of heat by radiation from the surface. This leads to a negative net radiation flux, *i.e.*, a loss of energy by the surface. When this happens, the sensible heat flux immediately above the ground surface changes sign, and the convection stops. In response to the radiational cooling of the surface, a ground-based stable boundary layer develops in which $\partial\Theta/\partial z > 0$ or $dT/dz > -\gamma_a$. Here, T is the temperature, $\Theta(z)$ is the mean potential temperature attained by an air parcel if it is compressed adiabatically from its initial state to a standard pressure of 1000 mb, and $\gamma_a = 0.01° \, C\,m^{-1}$ is the adiabatic lapse rate. As the thermal stratification in the SBL increases such that $dT/dz > 0$, it is often referred to as an inversion. In the SBL, turbulence is produced by friction at the ground surface, suppressed by the stable thermal stratification, and destroyed by viscous dissipation. Turbulent kinetic energy (TKE) is maximum near the ground and decreases monotonically with increasing height. Because of the loss of kinetic energy to the negative buoyancy flux throughout its depth, turbulence in the SBL is considerably weaker than in the daytime convective boundary layer (CBL).

A characteristic feature of the SBL is a low-level jet or velocity maximum. The jet can result from different mechanisms such as synoptic-scale weather patterns, sloping terrain, mountain and valley winds, land and sea breezes, and inertial oscillations. The reduction of momentum flux is most pronounced near the top of the SBL. Figure 1.1 shows averaged profiles computed by Mahrt et al. (1979) from 30 soundings during the O'Neill field study (Lettau & Davidson, 1957). The velocity jet occurs above the turbulent layer as inferred by the Richardson number profile, $Ri\,(z)$. Cooling is the strongest in the surface layer, and is due to turbulent transfer of heat to the ground surface. In the upper thicker region of the SBL, cooling occurs due to clear-air radiation and this cooling can extend upwards to considerable height. The profiles in Fig. 1.1 are averaged over several nights; profiles for individual nights may differ substantially.

The boundary layer depth at night is about an order of magnitude smaller than its depth during the day. Turbulence is much weaker and vertical transfer is much slower than in the CBL. Data scatter, intermittency, and nonstationarity, are among the problems encountered during measurements in the SBL. Additional complications include the existence of thin turbulent layers, separated from the main turbulent region by laminar layers, strong local terrain effects, and occurrence of gravity waves superimposed on the turbulence. Thus, the stable boundary layer is inherently more complex than the CBL, and our understanding of its structure is limited by the difficulties in measurements and numerical simulations. In the CBL, the buoyant production of turbulence dominates all other sources of TKE, and its strong mixing action tends to homogenize the flow in the horizontal direction. For example, the effects of surface features with scale heights on the order of 100 m are restricted to downwind distances of only a few scale heights during the day. During the night, however, these effects can extend many kilometers downwind. In the absence of strong mixing, the boundary layer is free to respond to other influences such as radiation, gravity waves, katabatic winds, nonsteady conditions, and surface inhomogeneities. In many cases, all or some of these effects will be present, and their separation is difficult.

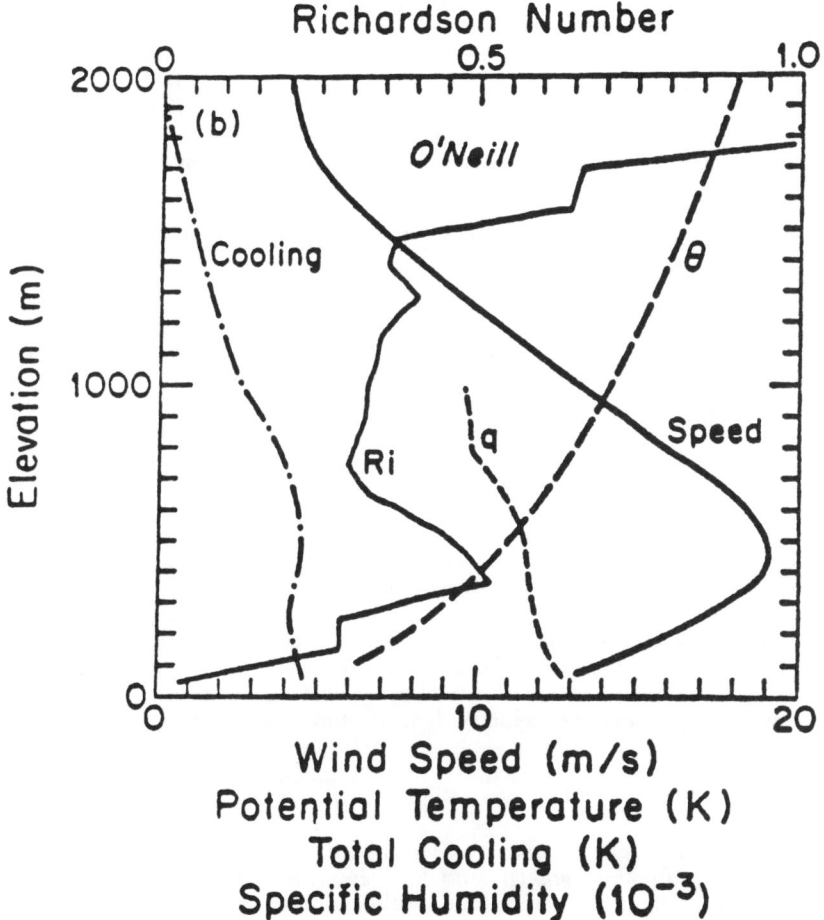

Figure 1.1: Averaged vertical profiles computed from 30 soundings of O'Neill data; the cooling profile shown was computed by subtracting the 0435 LST sounding from the 2035 LST sounding, both averaged over 5 days. The origin for potential temperature is 293 K. From Mahrt *et al.* (1979): reprinted by permission of Kluwer Academic Publishers.

In this chapter, selected features of the stable boundary layer are discussed in terms of similarity theories, observations, and numerical model results. Section 2 outlines the similarity theories that describe the vertical structure of the idealized stable boundary layer, including the mean profiles. Section 3 presents the SBL turbulence structure in terms of the similarity theories. Section 4 describes the general characteristics of the SBL and its key parameterizations, and discusses the role of radiative transfer and factors complicating the measurements in SBL, such as nonstationarity, intermittency, and gravity-waves. Section 5 summarizes the wide range of models used in the numerical simulation of flow and turbulence in the idealized SBL. Section 6 describes the Gaussian plume and Lagrangian particle diffusion models, and discusses the diffusion under nighttime stagnation conditions, using some recent research results. Section 7 presents the conclusions.

2 Vertical structure of the idealized SBL

2.1 Similarity theory

Similarity theory seeks to find empirical relations for mean and turbulence variables in the idealized boundary layer, which are made dimensionless by using characteristic scales of height, velocity, temperature, etc. A significant wind is required in the boundary layer, since similarity theory is not valid under nearly calm conditions. In the surface layer, which consists of the lowest 10 percent of the turbulent boundary layer depth h, the mean wind shear $(\partial U/\partial z)$ and potential temperature gradient $(\partial \Theta/\partial z)$ play an important role. In this layer, the turbulent fluxes are assumed to be essentially constant at their surface values. The Monin-Obukhov (M-O) similarity hypothesis postulates that the key parameters that control turbulent transfer in this layer are the height (z), buoyancy parameter (g/Θ_r), and kinematic turbulent shear stress (τ_o/ρ) and sensible heat flux $(\overline{w\theta_o})$ at the surface, where w is the fluctuating vertical velocity component, and Θ_r is a reference absolute potential temperature. From these parameters, velocity, length, and temperature scales can be defined as

$$u_* = (\tau_o/\rho)^{1/2}, \quad L = -u_*^3 \, \Theta_r/(k \, g \, \overline{w\theta_o}), \quad T_* = -\overline{w\theta_o}/u_* \, . \quad (2.1.1)$$

Here, u_* is friction velocity at the surface, T_* is scaling temperature, L is Monin-Obukhov length, and k is the von Kármán constant. The $(g/\Theta_r)\,\overline{w\theta_o}$ in the definition of L is the flux of buoyant acceleration due to density variations associated with turbulent fluctuations of temperature; when the surface is wet and moisture flux is large, a correction is added for density variations due to the lower molecular weight of water vapor compared to dry air. For unstable (daytime) conditions, $\overline{w\theta_o} > 0$ and $L < 0$; for stable (nighttime) conditions, $\overline{w\theta_o} < 0$ and $L > 0$; and for neutral conditions, both $\overline{w\theta_o}$ and $1/L$ approach zero. According to the M-O similarity theory, the surface layer variables, when nondimensionalized by the similarity scales L, u_*, and T_* should be universal functions only of the dimensionless height, z/L. This hypothesis was supported by the surface-layer data from Kansas, Wangara, and other experiments (see, e.g., Haugen, 1973; Hess et al., 1981; Panofsky & Dutton, 1984, Garratt & Hicks, 1990).

Nieuwstadt (1984) showed that, above the surface layer ($z > 0.1h$), the turbulence structure in the SBL follows "local" scaling, with $\tau(z)/\rho$, $\overline{w\theta}(z), g/\Theta_r$, and z as the controlling parameters. From these, the following similarity scales can be derived:

$$U_\ell = (\tau/\rho)^{1/2}, \quad \Lambda = -U_\ell^3 \Theta_r/(k\,g\,\overline{w\theta}), \quad T_\ell = -\overline{w\theta}/U_\ell . \tag{2.1.2}$$

According to local scaling hypothesis, SBL variables appropriately nondimensionalized with Λ, U_ℓ, and T_ℓ are functions only of z/Λ. In the so-called "z-less" (Wyngaard, 1973) or height-independent stratification layer ($\Lambda \le z \le h$) where turbulence parameters are insensitive to height, dimensionless variables formed with the above scales should be constants. The Minnesota (Izumi & Caughey, 1976), Cabauw (Nieuwstadt, 1984), and SESAME (Lenschow et al., 1988) data follow these local scaling predictions well, as shown later. Figure 2.1 depicts the various scaling regions in the idealized SBL discussed by Holtslag & Nieuwstadt (1986). The proximity of the $z/L = 1$ curve (dashed line) to the $z/\Lambda = 1$ curve in Fig. 2.1 suggests that $\Lambda \approx L$ in a large part of the SBL. This might explain why the surface layer results from the M-O similarity are often valid up to $z/L \approx 1$ (Wyngaard, 1973).

2.2 Mean wind and temperature profiles

Figure 2.2 shows the vertical profiles of mean wind and temperature in the SBL from the Minnesota experiment (Caughey et al., 1979). These profiles show large gradients through the SBL depth h of about 100 m. The wind profile shows the presence of a low-level nocturnal jet near the top of the SBL. The potential temperature profile shows a triple structure typical of a SBL in which turbulent cooling is more important than radiative cooling. Such nocturnal temperature profiles were also measured by an instrumented 200 m tower at Cabauw (van Ulden & Holtslag, 1985) in the Netherlands for moderate to strong winds.

In a horizontally homogeneous equilibrium surface layer, the nondimensional mean wind shear ϕ_m and mean potential temperature gradient ϕ_h follow the M-O similarity theory as follows:

$$\phi_m(z/L) = (k\,z/u_*)\,\partial U/\partial z = 1 + \beta_m z/L \tag{2.2.1}$$

$$\phi_h(z/L) = (k\,z/T_*)\,\partial \Theta/\partial z = Pr_t + \beta_h z/L , \tag{2.2.2}$$

where β_m and β_h are empirical constants, and Pr_t is the turbulent Prandtl number. The log-linear vertical profiles of mean velocity and temperature can be obtained by integrating the above equations:

$$U(z) = (u_*/k)\,[ln\,(z/z_o) + \beta_m z/L] \tag{2.2.3}$$

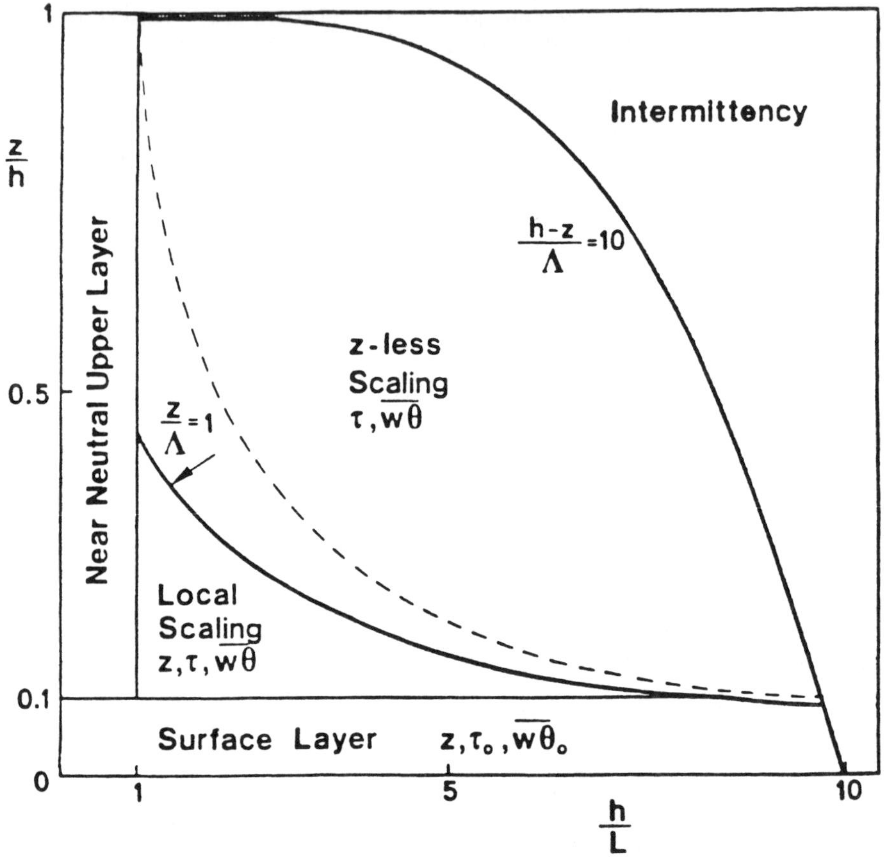

Figure 2.1: Scaling regions of the idealized SBL. The dashed line is given by $z/L = 1$ (see text). From Holtslag and Nieuwstadt (1986): reprinted by permission of Kluwer Academic Publishers.

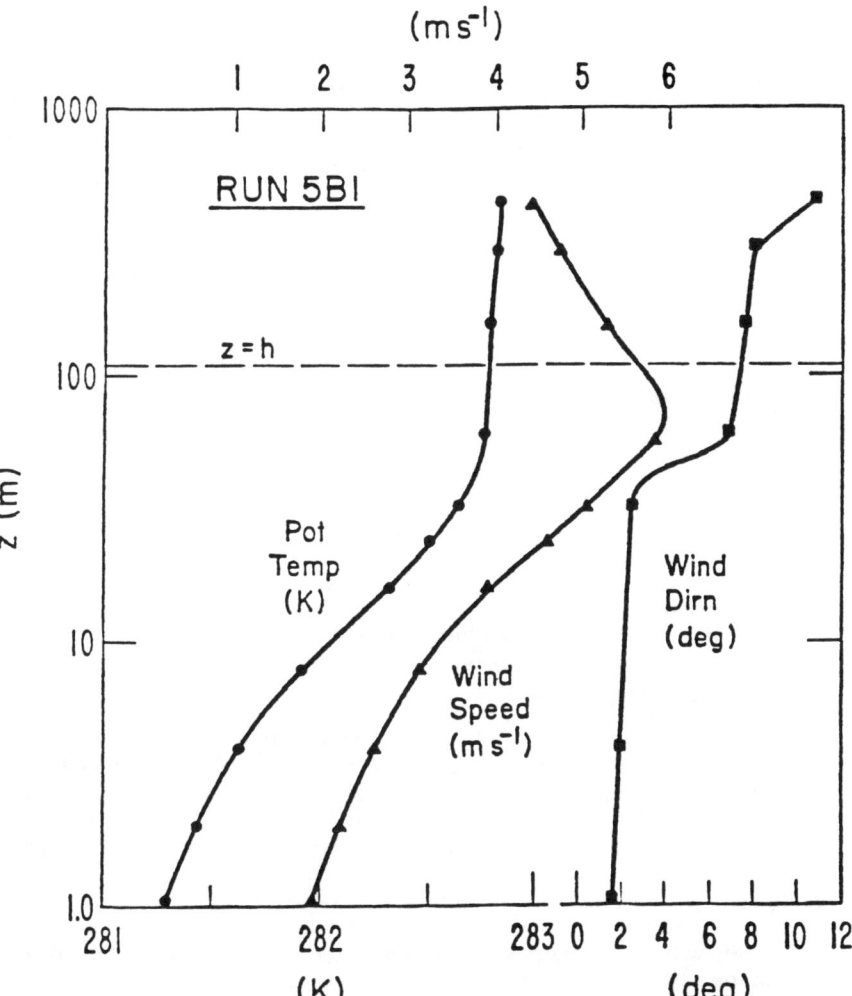

Figure 2.2: Vertical profiles of mean wind speed and direction, and potential temperature observed in the Minnesota experiment. From Caughey et al. (1979).

$$[\Theta(z) - \Theta_o] = (T_*/k) [Pr_t \, ln \, (z/z_o) + \beta_h \, z/L] \,, \tag{2.2.4}$$

where z_o is the surface aerodynamic roughness length defined such that $U(z_o) = 0$, and Θ_o is the potential temperature at the surface. From the 1968 surface layer experiment in Kansas, Businger $et\ al.$ (1971) derived $k = 0.35, Pr_t = 0.74$, and $\beta_m = \beta_h = 4.7$ for $0 \leq z/L \leq 1.0$. However, many differing sets of values of these parameters were reported since then by experiments at other sites. Högström (1996) critically reviewed these data and recommended the values $k = 0.40$, $Pr_t = 0.95$, $\beta_m = 5.3$, and $\beta_h = 8.0$ in the range $0 \leq z/L \leq 0.5$. For $z/L > 0.5$, most studies show increasing scatter, and both ϕ_m and ϕ_h level off and become independent of z/L.

Above the surface layer, local scaling implies that the nondimensional gradients of mean velocity and temperature approach constants ϕ_m^* and ϕ_h^* in the z-less state:

$$(k \, \Lambda/U_\ell) \, \partial U/\partial z \;=\; \phi_m^* \,, \quad (k \, \Lambda/T_\ell) \, \partial \Theta/\partial z \;=\; \phi_h^* \,. \tag{2.2.5}$$

Sorbjan (1987) presented SBL data from the Boulder Atmospheric Observatory that appear to support these predictions with some scatter, and suggested that the local similarity predictions are applicable only to ensemble-averages. The latter can be obtained by averaging many individual runs. In any case, these predictions need to be tested further with suitable data. The vertical profiles of scalars such as moisture and trace gases (from surface area sources) follow expressions similar to that given for the temperature.

We have noted that, in the absence of the strong vertical mixing action of the convective turbulence, the SBL is free to respond to other influences such as gravity-driven flows, terrain-induced perturbations, and synoptic scale pressure gradients and fronts. The low-level jet (LLJ) is a characteristic feature of the SBL. Kraus $et\ al.$ (1985) list some of the causes of the LLJ as follows: synoptic-scale baroclinity associated with weather patterns; baroclinity associated with sloping terrain, $i.e.$, drainage winds; fronts; advective accelerations; splitting, ducting, and confluence around mountain barriers; land and sea breezes; mountain and valley winds; and inertial oscillations. Blackadar (1957) proposed the inertial oscillation as a means of explaining the LLJ. When the boundary layer transitions from convective to stable conditions, the eddy diffusivity can decrease from 1 to 10^{-3} m^{2}s^{-1}. When the turbulent drag force is removed, the boundary layer flow becomes unbalanced and accelerates toward the geostrophic wind direction, leading to the development of a super-geostrophic wind speed maximum, known as the nocturnal jet, at the top of the nighttime boundary layer.

Katabatic winds can form over sloping terrains when the near-surface air cools, becomes negatively buoyant, and accelerates downslope. Katabatic winds, also known as drainage winds, slope or gravity flows, can exist over even gentle slopes of the order 0.001 radians (Brost & Wyngaard, 1978). Since there are relatively few such flat places on the Earth's surface, we can expect that katabatic winds will be a common feature in the SBL. The strong katabatic winds which frequently blow from the polar plateau toward the sea are a dominant surface phenomenon of the Antarctic climate (see, *e.g.*, Parish, 1988; André *et al.*, 1993).

3 SBL turbulence structure

3.1 Momentum, heat and other scalar fluxes

From governing equations for mean wind and temperature in a steady barotropic SBL over a flat terrain, it can be shown (Nieuwstadt, 1984; Lenschow *et al.*, 1988) that the vertical profiles of shear stress and heat flux are given by

$$\tau/(\rho u_*^2) = (1 - z/h)^{\alpha_1} , \quad \overline{w\theta}/\overline{w\theta}_o = (1 - z/h)^{\alpha_2} , \qquad (3.1.1)$$

where α_1 and α_2 are constants which depend on the temporal development of the SBL, terrain slope, advection effects, and other factors. Their values range from $\alpha_1 = 2$ and $\alpha_2 = 3$ in the evolving SBL in Minnesota (Caughey *et al.*, 1979) to $\alpha_1 \simeq 1.5$ and $\alpha_2 \simeq 1$ in the developed SBL at Cabauw (Nieuwstadt, 1984). The SESAME data (Lenschow *et al.*, 1988) indicate that $\alpha_1 = 1.75$ and $\alpha_2 = 1.5$, which are supported by recent data from several sites in Sweden (Smedman, 1991), as shown in Fig. 3.1. Sorbjan (1986, 1989) argued that α_2 should not be smaller than α_1 in order to avoid a singularity in the mean temperature gradient at $z = h$. However, it is possible that formulations based on the local scaling hypothesis are not applicable at the top of the SBL.

The vertical profile of a scalar concentration (*e.g.*, water vapor) turbulent flux can also be expressed in a similar way as

$$\overline{wc}/\overline{wc_o} = (1 - z/h)^{\alpha_3} . \qquad (3.1.2)$$

From Eqs. (2.1.1), (2.1.2), and (3.1.1), we obtain the relations:

$$U_\ell = u_* \left(1 - z/h\right)^{\alpha_1'} , \quad T_\ell = T_* \left(1 - z/h\right)^{(\alpha_2 - \alpha_1')} ,$$

$$\Lambda = L \left(1 - z/h\right)^{(3\alpha_1' - \alpha_2)} , \quad \alpha_1' = \alpha_1/2 . \qquad (3.1.3)$$

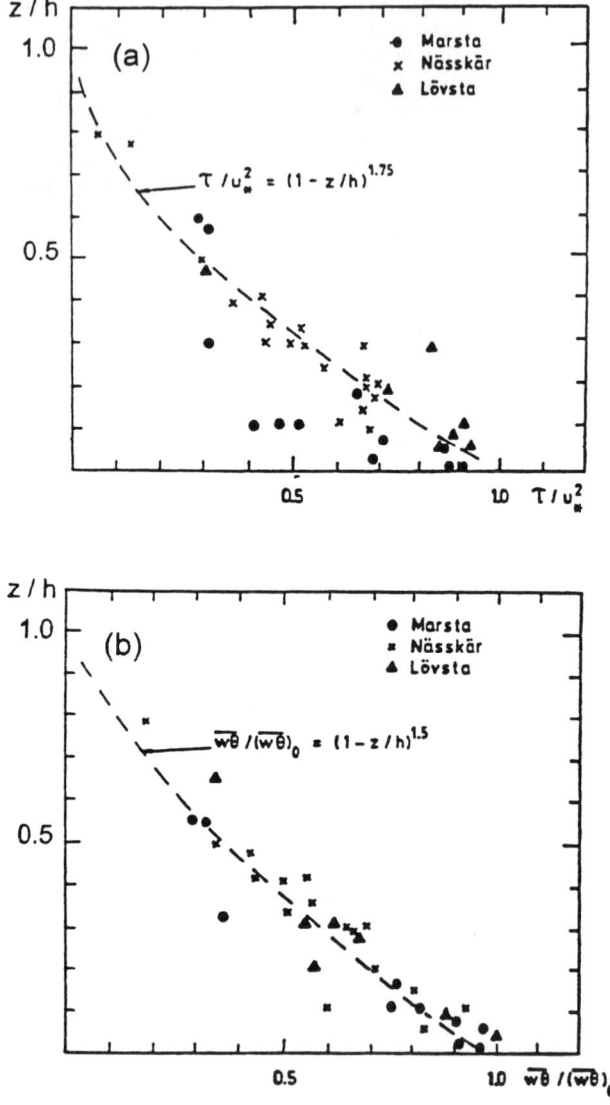

Figure 3.1: Vertical profiles of turbulent (a) shear stress and (b) sensible heat flux in SBL using data from 3 sites in Sweden. Dashed line shows fit to SESAME data given by Lenschow *et al.* (1988). From Smedman (1991).

3.2 Variances of turbulence

In the z-less stratification layer ($\Lambda \le z \le h$), turbulence quantities nondimensionlized with Λ, U_ℓ, and T_ℓ should be constants. For example, the velocity variances are given by $\overline{u_i^2}/U_\ell^2 = C_i$, where $C_i (i = 1, 2, 3)$ are constants. Using Eq. (3.1.3), this can be expressed as

$$\overline{u_i^2}/u_*^2 = C_i \left(1 - z/h\right)^{\alpha_1} , \qquad (3.2.1)$$

and the turbulent kinetic energy, $\bar{e} = 0.5 \left(\overline{u^2} + \overline{v^2} + \overline{w^2}\right)$, is given by

$$\bar{e}/u_*^2 = C_e \left(1 - z/h\right)^{\alpha_1} . \qquad (3.2.2)$$

The SBL velocity variances and TKE measured in SESAME (Lenschow et al., 1988) fit the above expressions fairly well with the values $\alpha_1 = 1.75, C_1 = C_2 = 4.5, C_3 = 3.1$, and $C_e = 6$, as shown in Fig. 3.2. Similarly, local scaling of the temperature variance gives $\overline{\theta^2}/T_\ell^2 = C_t$ which, using Eq. (3.1.3), leads to

$$\overline{\theta^2} = C_t \left(1 - z/h\right)^{(2\alpha_2 - \alpha_1)} . \qquad (3.2.3)$$

With $\alpha_2 = 1.5$, this equation fits SESAME data (see Lenschow et al., 1988), but with more scatter than for the velocity variance.

Sorbjan (1986, 1989) demonstrated that Eqs. (3.2.1) to (3.2.3) fit the Minnesota data (Izumi & Caughey, 1976; Caughey et al., 1979) well with $\alpha_1 = 2$ and $\alpha_2 = 3$. He derived the functional forms for many other turbulence quantities from the local scaling hypothesis, and showed that they also fit the Minnesota data; however, these parameterizations need to be tested with other independent data sets.

3.3 Turbulent kinetic energy and Richardson number

The equation for turbulent kinetic energy, $\bar{e}$, in a horizontally homogeneous SBL can be written as

$$\partial \bar{e}/\partial t = SP + BP - \partial(\overline{wp}/\rho)/\partial z - \partial \overline{we}/\partial z - \epsilon . \qquad (3.3.1)$$

Here, $SP = -\overline{uw}\,\partial U/\partial z - \overline{vw}\,\partial V/\partial z$ denotes the shear production of TKE, and $BP = (g/\Theta_r)\overline{w\theta}$ is the buoyant production, representing the destruction of TKE by negative buoyancy in the SBL; U and V are horizontal mean velocity components in x and y directions, respectively. Note that the ratio of the buoyant production term divided by the shear production becomes $-z/L$ near the ground. The last two terms in Eq. (3.3.1) are turbulent transport and viscous dissipation rate of TKE, respectively. The pressure transport (third term), usually calculated as a residual, acts to redistribute energy.

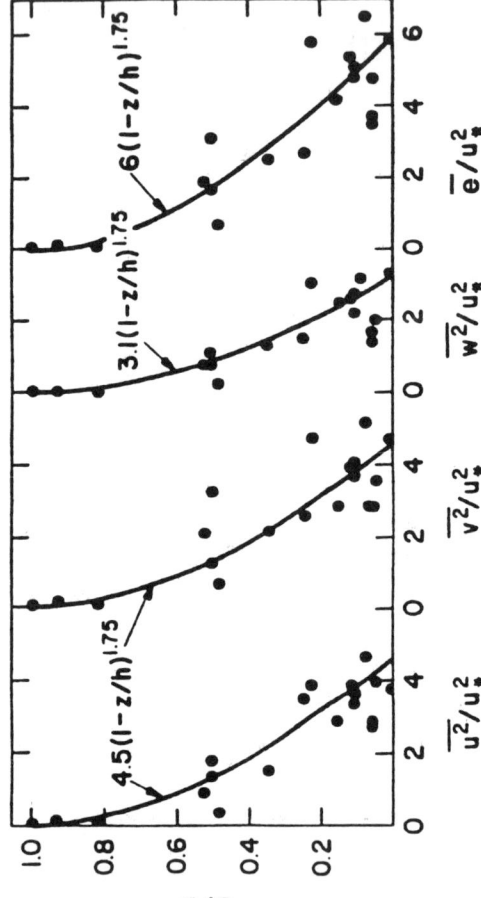

Figure 3.2: Vertical profiles of turbulent velocity variances and TKE observed in SESAME. The solid lines show the best-fit equations. From Lenschow *et al.* (1988); reprinted by permission of Kluwer Academic Publishers.

The components of the TKE budget in the SBL, derived by Lenschow *et al.* (1988) from aircraft data in the 1979 SESAME experiment in Oklahoma over a terrain with an estimated mean slope of 0.003, are shown in Fig. 3.3. The dominant terms are shear production and viscous dissipation, with a significant residual due to pressure transport, which is negligible after integrating over the entire boundary layer. The buoyancy term is relatively small and represents a loss. This TKE budget, with small buoyancy and large shear production terms near the surface, indicates that L was large.

The magnitudes of the shear production and buoyant destruction terms in Eq. (3.3.1) play an important role in the maintenance of turbulence and its intensity. The ratio of $-BP$ to SP gives the flux Richardson number (R_f):

$$R_f = -BP/SP . \qquad (3.3.2)$$

Richardson (see Businger, 1973) postulated that turbulence cannot be maintained when R_f reaches a critical value of 1, because the shear production is entirely compensated by the buoyant destruction of TKE. A gradient-Richardson number Ri, defined as

$$Ri = (g/\Theta_r)(\partial\Theta/\partial z)/[(\partial U/\partial z)^2 + (\partial V/\partial z)^2] \qquad (3.3.3)$$

is more widely used as a stability parameter in the SBL. Both Ri and R_f are positive in stable flows and are related as

$$Ri = (K_m/K_h)R_f , \qquad (3.3.4)$$

where K_m and K_h are the eddy diffusivities of momentum and heat, respectively. Assuming $K_m = K_h$, the critical value of Ri for the termination of turbulence is 1. This value represents an upper limit, however, and turbulence will be terminated even before Ri reaches a value of 1. A laminar flow can become unstable as perturbations grow and a breakdown into turbulence (onset) occurs when Ri is smaller than a critical value, $Ri_c = 1/4$, as deduced by Miles (1961).

Businger (1973) provided a qualitative description of the sequence of events leading to this transition and the consequences of reaching Ri_c in the SBL. As the ground cools at night, the establishment of a stable thermal stratification near the ground causes the Richardson number to approach and finally exceed its critical value. Turbulence is then supressed, a laminar layer develops, and the downward transfer of heat and momentum from the higher levels is impeded. Hence, the wind near the ground surface diminishes and a near-surface calm develops. Momentum is still transferred downward above the laminar layer. The winds aloft are essentially decoupled from the surface, and the upper-level air is free to accelerate with little resistance from surface friction. A strong wind shear then builds up, and because there is not a similar increase in the sensible heat flux, Ri must decrease and eventually become less than its critical value. This leads to the destruction of the laminar layer

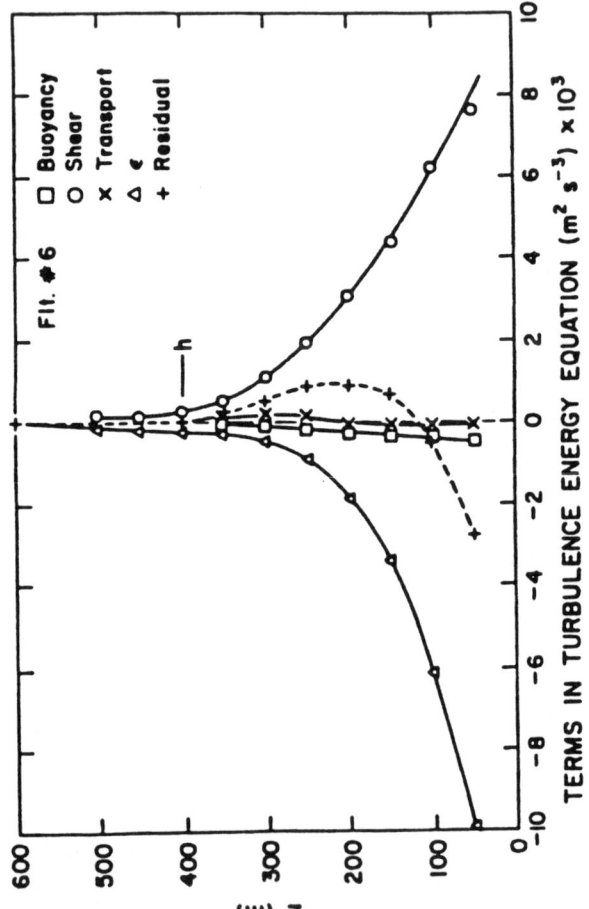

Figure 3.3: Vertical variation of TKE budget terms derived from SESAME data. From Lenschow et al. (1988): reprinted by permission of Kluwer Academic Publishers.

from above by turbulence. The latter then diffuses downward, eventually reaching the ground surface as a burst of heat and momentum. The surface layer turbulence dissipates with time, and the above process is repeated. Because of the repetitive transitions from turbulent to laminar flow and vice versa, it is unlikely that long periods of steady state occur in the SBL. The constant flux layer may become quite small and disappear altogether. When Ri first exceeds its critical value at an elevated level z, the latter can develop into a capping layer. Briggs (1996) suggests this height is proportional to $(u_*^3/C)^{1/2}$, where $C = -(g/\Theta_r)\,d\Theta_o/dt$.

3.4 Spectral characteristics

Spectral analysis determines the distribution of turbulent kinetic energy, variance, or covariance of the atmospheric variables as a function of frequency or wavenumber. Analysis of spectra provides not only a direct test of similarity predictions, but also an indirect means of evaluating variances and covariances, and is helpful in designing field experiments. For example, the appropriate instrument response and averaging times can be determined as functions of stability and height above the ground surface through spectral analysis. Variances and covariances can be estimated by integrating the spectral density, *i.e.*,

$$\sigma_\alpha^2 = \int_n S_\alpha(n)\,dn \qquad (3.4.1)$$

and

$$\overline{\alpha'\beta'} = \int_n C_{\alpha\beta}(n)\,dn \ , \qquad (3.4.2)$$

where $S_\alpha(n)$ is the spectral density of variable α, $C_{\alpha\beta}(n)$ is the cospectrum of variables α and β, and n is the cyclic frequency. Figure 3.4 shows a schematic of an idealized energy spectrum. The spectrum can be divided into three general regions: an *energy-containing region* which contains the bulk of the turbulent energy produced by buoyancy and shear; an *inertial subrange* where energy is neither produced nor destroyed, but is transferred to smaller and smaller scales; and a *dissipation range* where kinetic energy is converted into internal energy.

Dimensional analysis shows that the one-dimensional spectrum for a velocity component in the inertial subrange is given by

$$S_\alpha(\kappa) = a\,\epsilon^{2/3}\,\kappa^{-5/3} \ , \qquad (3.4.3)$$

where $\kappa = 2\pi/\lambda$ is the wavenumber, λ is the wave length approximated by U/n in the mean wind direction, and a is the Kolmogorov constant. Estimates of a range between 0.5 and 0.6 for $\alpha = u$; for v and w, a is greater by a factor of 4/3 due to local isotropy (see, *e.g.*, Kaimal & Finnigan, 1994). Kaimal (1973) discussed the turbulence spectra, length scales, and structure parameters in the stable surface layer. Equation (3.4.3) can be used to calculate the eddy dissipation rate ϵ. Smedman (1988)

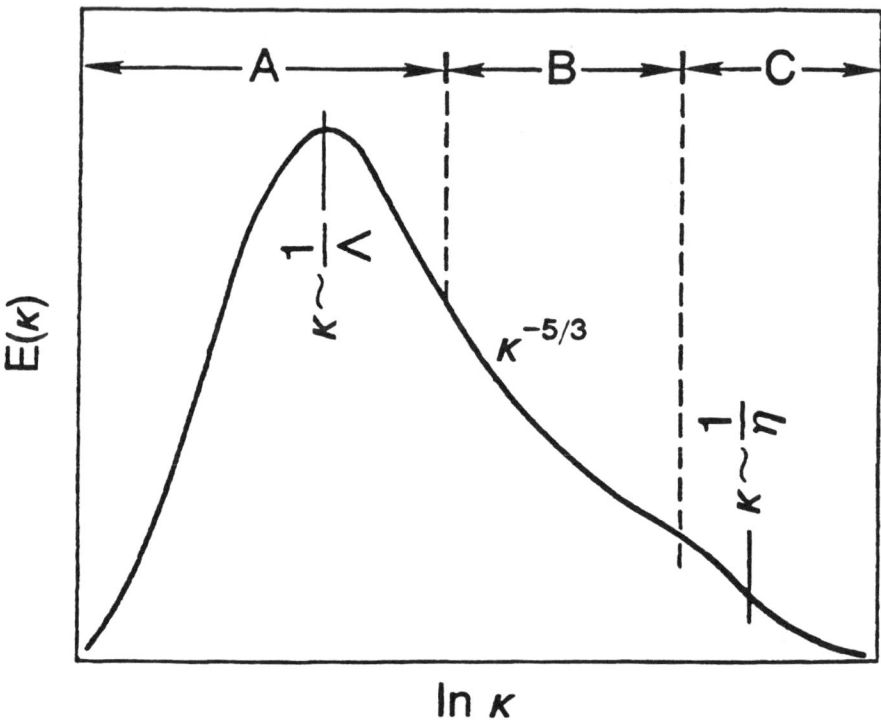

Figure 3.4: Idealized energy spectrum in the atmospheric boundary layer showing regions of energy production (A) and dissipation (C), and the inertial subrange (B). Here, Λ denotes the integral scale of turbulence and η the Kolmogorov microscale. From Kaimal and Finnigan (1994).

compared ϵ calculated from the w-spectra with ϵ derived from the TKE budget based on observations in the SBL. In the surface layer (at $z=2$m), there is a marked tendency for ϵ values calculated from the energy budget to exceed the values obtained from the spectra by about 30 percent. This can be explained by the difficulty in determining the inertial subrange in the w-spectrum at low heights, and the neglect of the transport terms in the TKE budget. However, the agreement is rather good in the upper SBL, indicating that the flow is in local equilibrium and turbulence is more isotropic in that region.

The study of spectra of turbulence components of atmospheric variables in the SBL is complicated by the presence of gravity waves and intermittency. Since the time scale (*i.e.*, oscillation period) of the higher frequency gravity waves is similar to the time scale of turbulence in the energy-containing and lower frequency range, wave energy can be mistaken for turbulence energy. Finnigan *et al.* (1984), for example, reported that gravity waves can account for up to 20 percent of the measured variance. It is important therefore to be able to distinguish between wave and turbulence signals; however, the methodology for doing this is still uncertain (see, *e.g.*, Hunt *et al.*, 1985; Finnigan, 1988).

The energy spectrum in the SBL can be divided into three subregions (see *e.g.*, Stull, 1988), and similarity theory can be used to evaluate the spectrum when interacting gravity waves are present. For wavenumbers $\kappa < \kappa_s$, where κ_s represents the wavenumbers of the dominant gravity waves, $S(\kappa) \propto \kappa^{-1}$. For eddies with wavenumbers κ in the buoyancy subrange $(\kappa_s < k < \kappa_b)$, where $k_b = C_1 N^{3/2} \epsilon^{-1/2}$ is the buoyancy wavenumber, $S(\kappa) \propto N^3 \kappa^{-3}$. Here, C_1 is a positive constant and $N = [(g/\Theta_r) \partial\Theta/\partial z]^{1/2}$ is the Brunt-Väisälä frequency. κ_b is the inverse of the Ozmidov length scale (Hunt *et al.*, 1985), which is the critical length where buoyancy forces become important and act to remove TKE. In the buoyancy subrange, the eddies are quasi two-dimensional because the static stability suppresses vertical motions. For wavenumbers $\kappa > \kappa_b$, the eddies do not feel the static stability directly, and the motions become three-dimensional. This subregion constitutes the inertial subrange in which $S(\kappa) \propto \kappa^{-5/3}$.

A gap in SBL turbulence spectra separating the buoyancy and inertial subregions is often observed (Caughey, 1982; Finnigan *et al.*, 1984; Lenschow *et al.*, 1988; Smedman, 1991). The gap usually occurs near the buoyancy wavenumber, and is quite noticeable in the surface layer (Caughey, 1982). In this layer, the turbulence eddy size scales with z so that, with increasing height, the inertial subrange shifts toward lower wavenumbers. Also, since N tends to increase with height, the buoyancy subregion shifts toward higher wavenumbers. The result is that the gap separating turbulence from waves eventually closes, and then it is no longer possible to distinguish between them. However, SBL spectra calculated from the Minnesota data show a spectral gap for all heights below the SBL depth h (Caughey, 1982).

4 General characteristics of the SBL

4.1 SBL parameterizations

4.1.1 SBL depth

The nighttime boundary layer depth is not well defined and is far more difficult to measure and predict than in the daytime case. However, it is an important parameter for estimating atmospheric dispersion from pollution sources, and for parameterizing the boundary layer in general circulation models. The depth h_t of the continuously-turbulent mixing layer over land at night is often referred to as the boundary layer depth h; h_t is usually determined from acoustic sounder or meteorological tower measurements, or estimated from numerical models as the height where the TKE or turbulent heat flux drops to a small fraction of its surface value. A diagnostic relation for equilibrium boundary layer depth h_e in terms of surface layer parameters was given by Zilitinkevich (1972) as

$$h_e = d \ (u_* L/|f|)^{1/2} \ , \tag{4.1.1}$$

where d is a constant and f is the Coriolis parameter. According to this equation, the quasi-steady SBL depth is proportional to the geometric mean of the length scales $u_*/|f|$ and L, and is determined uniquely by the wind shear, earth's rotation, and atmospheric stability. The empirical values of d quoted in literature vary widely from 0.2 to 0.7. Modeling studies over horizontal terrain suggest that $d = 0.40$ (Garratt, 1992). The value of d is sensitive to model parameterizations, and depends on time after evening transition, surface cooling rate, terrain slope and angle of fall line relative to geostrophic wind, and other factors.

Other characteristic heights can be defined in the SBL. These include the depth of the ground-based thermal inversion, h_i, and the height of the velocity maximum or jet, h_u. The relations between these parameters have been examined by Mahrt et al. (1979), Arya (1981), and André & Mahrt (1982), among others. h_i typically grows from a value of less than 100 m near sunset to a height of 200 to 500 m by early next morning, in spite of the increasing stability. h_i is often much larger than the mixing layer depth h_t due to clear-air radiational cooling, and correlates poorly with h_t. Observations (Mahrt et al., 1979) show that h_u is generally greater than h_e but less than h_i. h_u, which approximately coincides with the height of the maximum in Ri, is a better measure of h under strongly stratified conditions with a well developed low-level jet (Arya, 1981).

Observations sometimes differ from the behavior in the idealized SBL. Figure 4.1 shows the diurnal variations of h_i, h_u, and h_t calculated from observations taken by Smedman (1988) at a site near Marsta, Sweden during the period 23–25 February, 1983. Inversion height and the height of the wind maximum were measured remotely with a Doppler sodar, and h_t was calculated from Eq. (4.1.1). In the idealized SBL, the low-level jet is thought to occur at the top of the layer with significant turbulence, h_t; however, Smedman's observations showed a very different SBL structure.

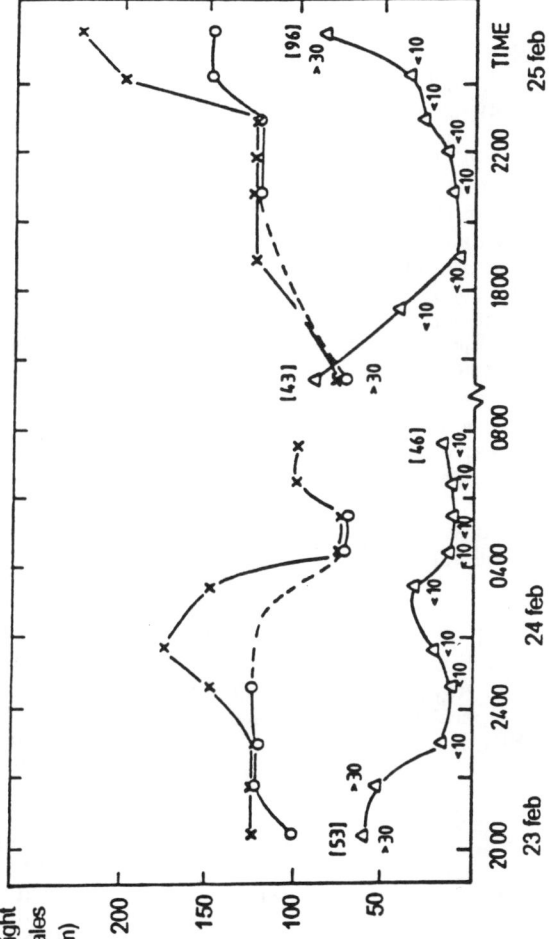

Figure 4.1: Diurnal variation of three scale heights: $\times = h_i$, $\circ = h_u$, and $\triangle = h_e$ calculated using Eq. (4.1.1) with $d = 0.4$. Numbers without brackets denote h_t values taken from Ri calculations, and numbers in brackets denote calculations using the stress profile, Eq. (3.1.1) with $\alpha_1 = 1.5$. From Smedman (1988): reprinted by permission of Kluwer Academic Publishers.

In the early evening, h_t extended up to about the level of the velocity jet, h_u; however, with continued radiational cooling, the surface layer stabilized and the turbulence at the higher levels died out. Later h_t was less than 10 m, but turbulence began to appear just below the velocity jet in response to the increased local wind shear. If h_t were based on sodar measurements, then this height would be taken to be the height of continuous turbulence; clearly, this was not the case. In fact, a two-level boundary layer developed: a surface-based stable turbulent layer, and an elevated less stable turbulent layer.

The SBL evolves slowly in time, and rarely reaches a steady state. This led Nieuwstadt & Tennekes (1981) to propose a linear prognostic rate equation for the evolving SBL depth h as

$$dh/dt = (h_e - h)/T_\theta ,   (4.1.2)$$

where the time scale $T_\theta = -(\Theta_h - \Theta_o)/\text{SCR}$ is a monotonically increasing function of time and is usually of the order of a few hours; $\text{SCR}=d\Theta_o/dt$ is the surface cooling rate and Θ_h is the mean potential temperature at the top of the boundary layer. Thus, h changes slowly in response to changes in external conditions, and will approach h_e a few hours after the evening transition. In middle latitudes, h values will be close to h_e.

Yu (1978) and Arya (1981), among others, evaluated several diagnostic and prognostic relations proposed for the nocturnal boundary layer depth and concluded that none of the diagnostic heights are significantly correlated with h_t under extremely stable conditions. Using the Cabauw data, Nieuwstadt (1984) showed that Eq. (4.1.1) predicts h_t well under moderately stable conditions. Derbyshire (1995) cautioned that predictions from this equation based on surface layer parameters tend to underestimate h_t over a locally smooth surface.

4.1.2 Geostrophic drag and heat transfer relations

The geostrophic drag and heat transfer laws provide a simple method for relating surface turbulent fluxes to large-scale, external flow parameters in PBL prediction models. Asymptotic matching of the similarity profiles of mean velocity and temperature in the surface and outer layers of the stationary, barotropic SBL leads to the relations (Zilitinkevich, 1975):

$$G \cos \alpha/u_* = [ln \{u_*/(|f|z_o)\} - A(\mu)]/k   (4.1.3)$$

$$G \sin \alpha/u_* = -sign(f) B(\mu)/k   (4.1.4)$$

$$(\Theta_h - \Theta_o)/T_* = Pr_t [ln \{u_*/(|f|z_o)\} - C(\mu)]/k ,   (4.1.5)$$

where $\mu = u_*/(|f| L)$, α is the angle between the geostrophic wind and the surface wind vectors, and A, B, and C are universal functions of μ to be determined from theory and observations. The first two equations may be combined to give an implicit relation between the geostrophic drag coefficient, $C_g = u_*/G$, and the surface Rossby number, $Ro = G/(|f|z_o)$, as follows:

$$ln\, R_o = A(\mu) - ln\, C_g + [(k/C_g)^2 - B^2(\mu)]^{1/2} \ . \qquad (4.1.6)$$

This implies that, for a given Rossby number, C_g should only depend on μ. The functions A, B, and C calculated by Rao & Snodgrass (1979) from a second-order closure model and Eqs. (4.1.3) to (4.1.5) agree reasonably well with the polynomial best-fits to the Wangara data (Clarke et al., 1971) given by Arya (1975), as shown in Fig. 4.2. For $\mu > 10$, the model results are given by the relations:

$$A(\mu) = ln\, \mu^{1/2} - 0.98\,\mu^{1/2} + 2.5 \qquad (4.1.6)$$

$$B(\mu) = 1.79\,\mu^{1/2} - 0.6 \qquad (4.1.7)$$

$$C(\mu) = ln\, \mu^{1/2} - 3\,\mu^{1/2} + 6.5 \ . \qquad (4.1.8)$$

There is extensive literature on the determination of the functions A and B; there is not as much information on C. The reader is referred to Zilitinkevich (1989) for a recent review. Garratt (1992) discussed the effects of baroclinity on the drag laws.

4.1.3 Length scales
Several length scales are used in the description of the structure of the SBL. Among them are $z, z_o, L, u_*/|f|$, and $|G/f|$ used in the surface and Rossby similarity theories. An important buoyancy length scale for turbulence in the SBL is given by

$$\ell_B = \sigma_w/N \ , \qquad (4.1.9)$$

where $\sigma_w^2 = \overline{w^2}$ and N is the Brunt-Väisälä frequency (§3.4). In a stably-stratified environment, the limiting size ℓ_B of the of the energy-containing eddies is determined by the turbulence velocity (σ_w) and the stratification $(\partial\Theta/\partial z)$. Assuming a balance between inertia forces $(\rho\sigma_w^2/\ell_B)$ and buoyancy forces $(\rho N^2 \ell_B)$ leads to Eq. (4.1.9).

A variety of formulations have been used for the turbulent mixing length ℓ_m in the SBL. Delage (1974), Nieuwstadt & Driedonks (1979), and Rao & Snodgrass (1981) used the expression:

$$\ell_m^{-1} = [1/(kz) + 1/\lambda + \beta_m/(kL')] \ , \qquad (4.1.10)$$

where $L'(z)$ is the local value of the M-O length scale defined analogous to Λ in Eq. (3.1.3), and $\lambda = 0.0004\,|G/f|$. Thus, the mixing length ℓ_m is proportional to z near the ground, and is limited by either λ (under near-neutral conditions) or kL'/β_m (under stable conditions). Brost & Wyngaard (1978) approximated the viscous dissipation rate as $\epsilon = d_1 q^3/\ell_1$, where $q = (2\bar{e})^{1/2}$ is the turbulent velocity, ℓ_1 is an integral length scale of the energy-containing eddies, and $d_1 = 0.139$; they specified ℓ_1 as

$$\ell_1^{-1} = (1/z) + 1/(c_1 \ell_B) \ , \qquad (4.1.11)$$

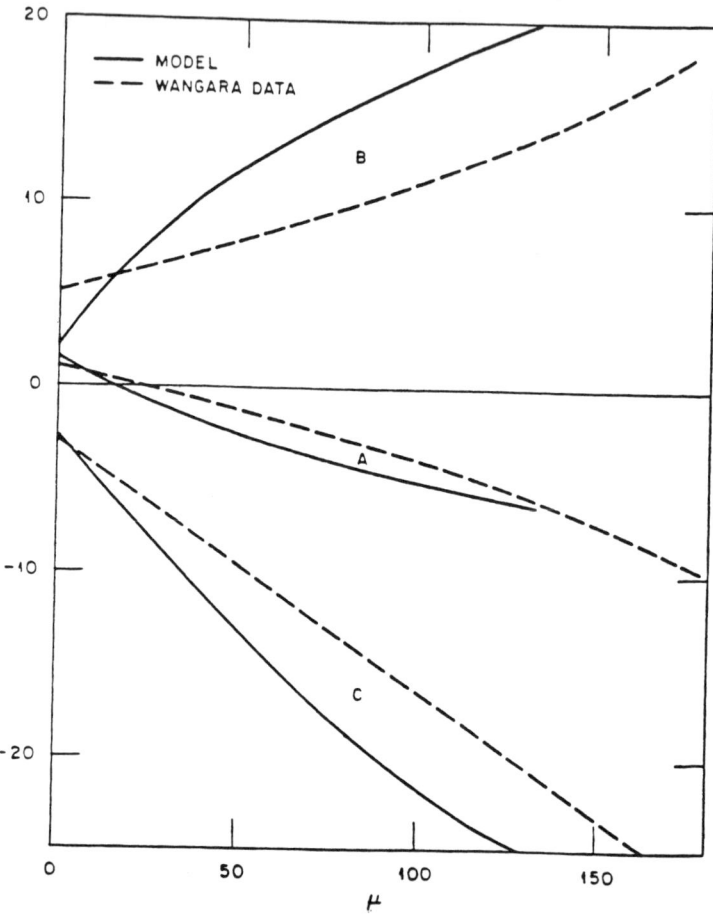

Figure 4.2: Comparison of similarity functions $A(\mu), B(\mu)$, and $C(\mu)$ computed from a second-order closure model to the polynomial best-fits to Wangara SBL data given by Arya (1975). From Rao & Snodgrass (1979): reprinted by permission of Kluwer Academic Publishers.

where $c_1 = 1.69$ was chosen to give a constant flux Richardson number. Equation (4.1.11) is based on the hypothesis that the largest vertical scale of the turbulent motions in stratified flows is determined by the smaller of the height z above the surface and the buoyancy scale ℓ_B. Alternately, Hunt et al. (1985) proposed $\epsilon = d_2 \sigma_w^3/\ell_2$, where d_2 varies from 0.4 to 0.6, and

$$\ell_2^{-1} = (A'/z) + (B'S)/\sigma_w . \tag{4.1.12}$$

Here, $S = [(\partial U/\partial z)^2 + (\partial V/\partial z)^2]^{1/2}$ is the magnitude of the mean shear, and $A' = 0.5$ and $B' = 0.7$ are empirical constants. Equation (4.1.12) thus employs a harmonic mean to emphasize the smaller of the inner ('blocking') and outer ('shear') scales. This formulation neglects the direct influence of stratification and assumes that the dissipation rate is set by the mean strain rate. This assumption is consistent with the fact that the shearing scale, σ_w/S, is typically smaller than the buoyancy scale σ_w/N in stratified turbulence. Note that $(N/S)^2 = Ri$, see Eq. (3.3.3), which is often of the order of $1/4$ through a large part of the upper SBL. This implies $S \approx 2N$ in this layer.

Other formulations of mixing length as a function of Richardson number have been used to directly relate turbulent shear stress to the mean wind shear. A comparative assessment of the mixing length formulations in the SBL was provided by Lacser & Arya (1986).

4.1.4 Eddy diffusivity profiles

For an idealized barotropic SBL, Businger & Arya (1974) specified the steady-state eddy diffusivity distribution by the equation

$$K_m f/u_*^2 = [k\xi/(1 + \beta_m \mu \xi)] \, exp\{-|V_g/u_*|\xi\} , \tag{4.1.12}$$

where $\xi = |f|z/u_*$, and $V_g = G \sin\alpha$. Brost & Wyngaard (1978) predicted the K_m profile from a second-order closure model as

$$K_m/(ku_* h) = Z(1-Z)^{1.5}/(1 + \beta_m Z h/L) , \tag{4.1.13}$$

where $Z = z/h$. Rao & Snodgrass (1979) also calculated K_m variation from a second-order closure model and expressed it in terms of the surface layer parameters as

$$K_m = (ku_* z/\phi_m) \, exp(-b\eta) , \tag{4.1.14}$$

where $\phi_m = 1 + \beta_m \zeta$, $\eta = \zeta\mu^{-1/2}$, $\zeta = z/L$, and $b = 9.1$. From M-O similarity theory (§2.2), the eddy diffusivity of heat can be obtained as

$$K_h = K_m(1 + \beta_m \zeta)/(Pr_t + \beta_h \zeta) . \tag{4.1.15}$$

4.2 Role of radiative transfer

The equation governing the rate of change of potential temperature can be written as

$$\partial \Theta / \partial t = -\partial \overline{w\theta} / \partial z + (\rho\, c_p)^{-1} \partial F_N / \partial z + R \,, \qquad (4.2.1)$$

where F_N is net longwave radiative flux, c_p is specific heat of air at constant pressure, and R is a residual accounting for temperature advection as well as errors in estimating the other terms in Eq. (4.2.1). The observed temperature changes thus result from a combination of turbulent and radiative flux divergences. The vertical profiles of radiative fluxes depend on surface temperature, as well as on profiles of temperature, carbon dioxide, water vapor, and ozone within the boundary layer and throughout the troposphere. Under clear skies and moderate or weak winds at night, the net radiative-flux divergence is an important cooling mechanism.

The SBL tends to evolve towards a stationary state in which the total cooling rate, $\partial \Theta / \partial t$, becomes invariant with height. Since the radiative cooling rate near the surface is much larger than the total cooling rate within the bulk of the boundary layer, the condition of constant total cooling with height may actually require turbulent warming (heat-flux convergence) near the surface. This leads to a maximum in the $\overline{w\theta}$ profile just above the surface, as demonstrated in the model results of Garratt & Brost (1981) and Tjemkes & Duynkerke (1989). However, available observations are not good enough either to infer the maximum in heat flux just above the surface or to verify the model results.

Based on the importance of radiative cooling relative to turbulent cooling, Garratt & Brost (1982) suggested a three-layer structure of the SBL. In the surface layer ($z/h < 0.1$) and near the top of the SBL ($0.8 < z/h < 1$), radiative cooling is important and the potential temperature profile shows a strong negative (convex) curvature. In the middle layer ($0.1 \leq z/h \leq 0.8$), turbulent cooling dominates and $\Theta(z)$ is nearly linear with a slight positive curvature ($\partial^2 \Theta / \partial z^2 > 0$). The radiative cooling rate is nearly independent of height above the surface layer. The temporal evolution of the bulk SBL parameters, as well as velocity, turbulent stress, and diffusivity profiles are little affected by radiative cooling. The latter mainly affects the mean temperature and heat flux profiles by decreasing the temperature gradient over the SBL, which therefore becomes less stable and much thicker compared to the case with no radiative cooling. Above the SBL ($z/h > 1$), cooling is by clear-air radiative effects alone, and large Ri values are generated.

Using mean profiles from Wangara (Australia) and Voves (France) experiments, André & Mahrt (1982) found that radiative cooling is primarily associated with moisture, though it is only modestly sensitive to the particular moisture distribution. The turbulent heat and radiation flux divergences make comparable contributions (about 40 percent each) to the heat budget over the entire inversion layer, with the residual R

accounting for the remaining 20 percent of the observed cooling. Clear-air radiative cooling extends the surface inversion to levels several times higher than the turbulent layer, *i.e.*, $h_i \gg h_t$. Based on this study, André & Mahrt suggested a two-layer model of the nocturnal surface inversion in which the lower layer is dominated by turbulence, and the upper layer by clear-air radiative cooling. However, many numerical models of the SBL neglected the radiative-flux divergence, and considered only radiative cooling at the surface and the resulting downward turbulent heat transfer to the cooled surface.

Estournel *et al.* (1986) calculated the vertical profiles of the terms of Eq. (4.2.1) from the data of the 1980 ECLATS experiment in Niger; these terms vary as depicted in Fig. 4.3. The radiative cooling shows little variation with altitude except in the lowest few meters. At the beginning of the night, the turbulent cooling rate decreases strongly with increasing height, which leads to the inversion formation. Though the vertical gradient of turbulent cooling decreases a few hours later, this term remains a major component of atmospheric cooling in the nocturnal boundary layer for moderate winds. The relative contributions of radiative and turbulent terms strongly depend on the geostrophic wind speed (Estournel & Guedalia, 1985). For strong winds, the depth of the stable layer overlying the turbulent layer shows little variation during the night. For weak winds, the contribution of radiative cooling becomes more important, especially at increasing heights from the surface. On a clear night, the typical depth of the surface inversion layer is about 100 m, and the sensible heat flux near the ground ranges from - 5 to - 30 W m^{-2} depending on the wind speed and cooling rate of the ground. The surface temperature decreases for the first few hours as square root of time from evening transition, and then decreases more slowly (Brost & Wyngaard, 1978; Surridge, 1986).

4.3 SBL turbulence and measurement problems

4.3.1 Averaging and the ergodic hypothesis
Theories of turbulence are based on the calculation of ensemble, time, and space averages. While turbulence theories are almost always based on the existence of ensemble averages, such averages are seldom calculated except in the laboratory. Meteorologists use the time average for observations taken at a fixed location, and the space average for observations taken on a moving platform. To agree with theory, the time and space averages must correspond to the ensemble average. The ergodic hypothesis states that these averages are equal when the turbulence is stationary and homogeneous. Stationarity requires the invariance of average quantities with time, and homogeneity requires the invariance of average quantities in space. In the atmospheric boundary layer, where

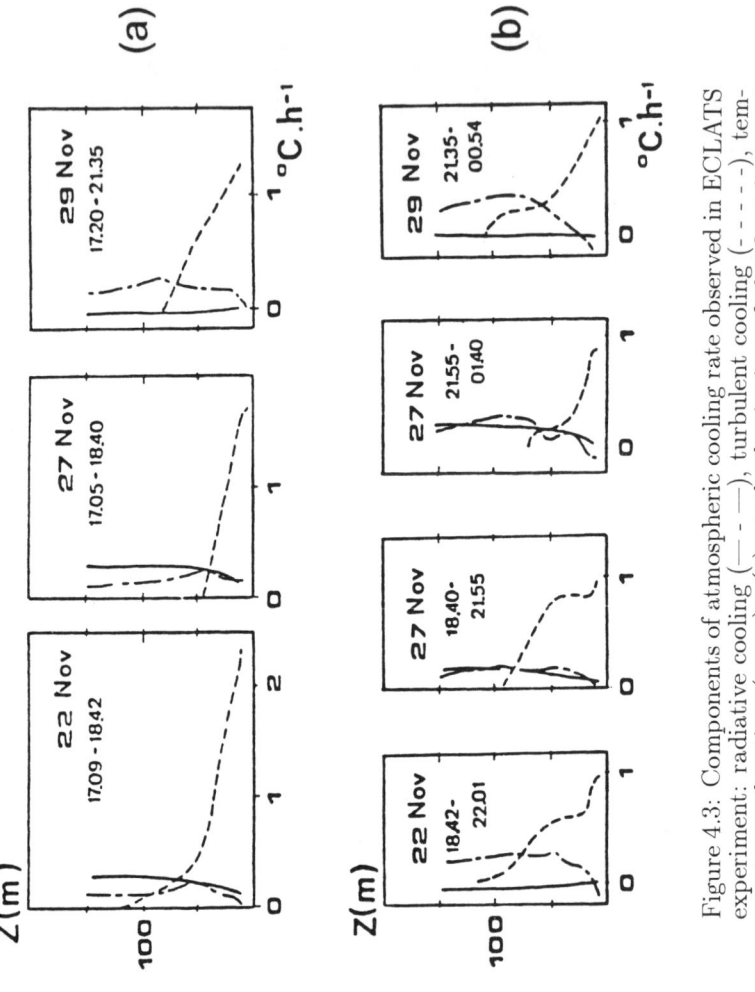

Figure 4.3: Components of atmospheric cooling rate observed in ECLATS experiment: radiative cooling (– · –), turbulent cooling (- - - -), temperature advection (———): (a) at the beginning of the 3 nights, and (b) for the following periods. From Estournel et al. (1986): reprinted by permission of Kluwer Academic Publishers.

the diurnal variations of mean flow and turbulence quantities are significant, stationarity may be approached for only a few hours or less. Consider the time average defined by

$$\bar{s}^T(x) = (1/T) \int_{T/2}^{T/2} s(x,t) \, dt \; . \tag{4.3.1}$$

The condition of stationarity is met if the averaging time $T \to \infty$; for real applications, however, it is required that $T \gg \tau_i$, where

$$\tau_i = \int_0^\infty R(t) \, dt \tag{4.3.2}$$

is the integral time scale of turbulence, and $R(t)$ is the autocorrelation coefficient. The integral time scale corresponds to the time scale for the largest turbulent eddies. In the convective boundary layer, this is similar to the convective time scale, $t_* = z_i/w_*$, where z_i is the depth of the mixed layer and w_* is the scale velocity for the convective updrafts. Thus, t_* is a measure of the time required for an air parcel to travel from the surface layer to the top of the mixed layer in a convective eddy; typically this is about 15-20 min. In the stable boundary layer, a similar definition is not immediately obvious because of the absence of unique height and velocity scales. Brost & Wyngaard (1978) suggested a time scale for diffusing the surface effects through the SBL as

$$t_* = h/(0.01 \, u_*) \; , \tag{4.3.3}$$

where h is the depth of the boundary layer with continuous turbulence. For typical values of h and u_*, this time ranges from 7 to 30 h (Stull, 1988). Thus, the calculation of time averages in the SBL, especially in the upper regions, remains a problem, and it is not certain that the ergodic hypothesis holds.

The use of aircraft for observing the boundary layer is becoming more popular following the development of fast and accurate global positioning systems. Examples of SBL analyses based on aircraft measurements include Mahrt (1985), Mahrt & Gamage (1987), Lenschow et al. (1988), and Kim & Mahrt (1992). Generally, aircraft observations are taken along a straight line, and the appropriate average is

$$\bar{s}^L(t) = (1/L) \int_{L/2}^{L/2} s(x,t) \, dx \; . \tag{4.3.4}$$

This average is similar to the time average, but now we require horizontal homogeneity instead of stationarity. If the flight segment L is sufficiently large, i.e., $L \gg \mathcal{L}_s$ where $\mathcal{L}_s$ is the integral length scale of the turbulence, then homogeneity is assumed. But little is known about $\mathcal{L}_s$, especially within the boundary layer. Terrain effects can produce standing flow perturbations and gravity waves, which will not be observed at a fixed location, but will appear to an aircraft as turbulence. The integral scale can become so contaminated by the mesoscale heterogeneity that, when one computes the integral scale, it is not necessarily the "turbulence integral scale." Also, horizontal gradients of mean quantities will appear as time trends in the data. This poses special problems in processing aircraft data. Thus, any turbulence calculation will be sensitive to averaging path length, detrending process, and filtering.

Generally, the structure of the SBL observed from fixed and moving platforms is different, and reconciling this difference remains a problem. Turbulence is a property of the flow, *i.e.*, it is essentially Lagrangian in character, and it is reasonable to assume that observing as much of the flow as possible is the best method. However, current theories and models are based on an Eulerian or fixed frame approach. In such a view, turbulence is local. If the ergodic hypothesis is satisfied, then the Lagrangian and Eulerian views are equivalent. However, ergodicity in the SBL, except in rare circumstances, is seldom realized.

4.3.2 Intermittency and breakdown of the SBL

It is recognized that turbulence in the very stable SBL is intermittent or burst-like (Mahrt, 1985; Nappo, 1991), and that patches of turbulence in various states of decay can be found throughout the SBL. Mahrt (1989) describes two possible forms of turbulence intermittency: "local" or small scale, and "global". Local intermittency is identified with the fine scale structure internal to the main eddies, and is associated with the dissipation of TKE. The sharp edges of the main eddies contribute to the intermittency due to generation of smaller scale turbulence by the eddy scale shear. In some cases, the small scale intermittency occurs as numerous narrow zones of shear with uncertain relationship to the main coherent structures. Global intermittency is identified with patches of turbulence and large intervening areas with little turbulence. Typically, the distance between these patches is large compared to the scale of the patches. One of the biggest problems in measuring surface-layer fluxes in the SBL is the question of sampling time in the presence of global intermittency. Sufficiently long records may include significant nonstationarity (Wyngaard, 1973), partly due to diurnal trends, and may include significant advection as well. Thus, in some cases it may not be possible to accurately measure fluxes at a fixed point. These sampling problems ultimately lead to large scatter when attempting to form similarity relationships.

In practice, turbulence in the SBL is assumed to be stationary. Under conditions of moderate stability, when the shear production is strong, stationarity can be observed in the surface layer. However, away from the ground surface in the upper regions of the SBL, turbulence is often intermittent and sporadic. Figure 4.4 shows a time series of wind speed–temperature covariation, $C_{UT}(t)$, superimposed on a sodar trace observed by Nappo (1991) atop a 60 m ridge near Oak Ridge, Tennessee. Here, C_{UT} is defined as

$$C_{UT}(t) = (U - < U >) (T - < T >) , \qquad (4.3.5)$$

where $U(t)$ is the horizontal wind speed, $T(t)$ is the air temperature, and <> denotes here a running mean. The quantity $-\rho \, c_p \, C_{UT}$ represents a horizontal heat flux due to turbulence; this flux was assumed to be related in a linear way to the vertical turbulent heat flux. Data for calculating $C_{UT}(t)$ were taken from the 40 m height level of a tower located about 400 m away from the sodar. This height level corresponds to a height of 100 m above the valley floor, and represents conditions in the upper SBL. It can be seen from Fig. 4.4 that the covariation is essentially zero until just before 2200 h when a sharp rise occurs. This jump in C_{UT} corresponds with the sudden appearance of turbulence as indicated by the sodar trace at the 40 m height. After this initial period, the covariation and the turbulence are seen to vary with time; however, a correlation exists between these quantities if examined along the 40 m height of the sodar trace.

The intermittency of turbulence seen in Figure 4.4 is typical for the upper SBL; however, on occasion the stable stratification through the entire SBL can break down. The cause-and-effect relations associated with these breakdowns are not yet clear. Businger (1973) gave a qualitative description of the breakdown process (see §3.3). Kondo et al. (1978) suggests that under very stable conditions, there exists a distinguishable interface dividing the ground-based active turbulent layer from a quiescent layer above. Undulations of this interface are produced by Kelvin-Helmholtz instability which develops in response to the speed and density differences across this interface. The movement of these undulations past an instrumented tower are marked by periods of enhanced turbulence and negative sensible heat flux. Data from Kondo et al. (1978) indicate a time scale of about ten minutes for these periods.

4.3.3 Gravity-wave generated turbulence

Gravity waves are often observed in the SBL (Einaudi et al., 1989), and the interaction of these waves with the mean flow often results in turbulence production (e.g., Nappo, 1992). However, these turbulence-producing mechanisms have not yet been accounted for in conventional boundary layer theory. Perhaps a reason for this is the difficulty in separating wave and turbulence signals. Bretherton (1969) suggested that it may be impossible to distinguish rigorously between waves and turbulence in the SBL because their time scales are similar. Chimonas (1972) proposed that the dynamic stability of a stable parallel shear flow can be modulated by the presence of a finite amplitude wave, and that turbulent breakdowns of the flow can occur if the stability is low. Theoretical and observational studies of the interactions of waves and turbulence in the SBL have been performed by, for example, Fua et al. (1982), Finnigan (1988), and Einaudi & Finnigan (1993).

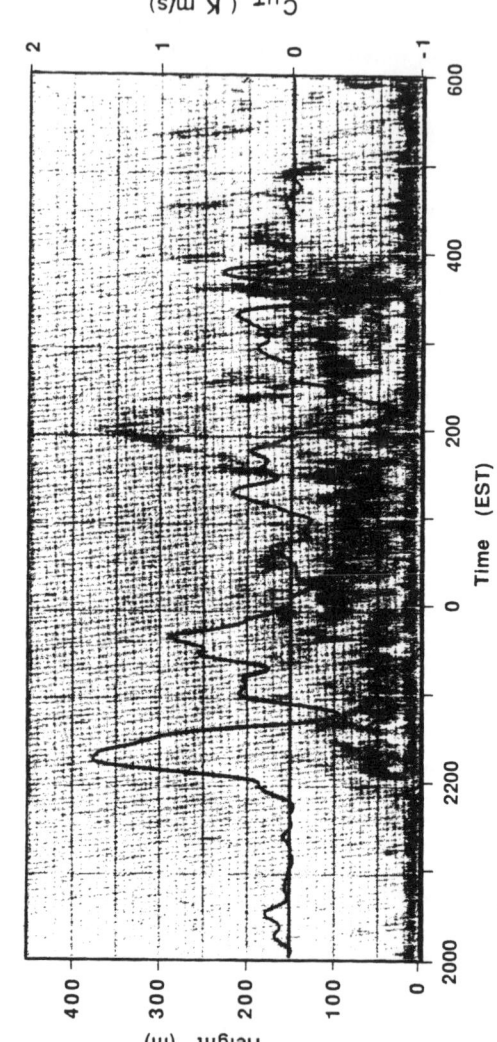

Figure 4.4: Wind speed–temperature covariation (C_{UT}), and sodar trace observed on a 60 m high ridge for a typical night near Oak Ridge, TN. Data for calculating C_{UT} were obtained at 40 m height on a nearby tower. From Nappo (1991): reprinted by permission of Kluwer Academic Publishers.

Small-scale terrain features can generate gravity waves in the SBL for typical values of wind speed and stratification. The stress on the SBL flow associated with terrain-generated gravity waves has been studied by Chimonas & Nappo (1989), and Nappo & Chimonas (1992). As these waves propagate away from the surface, they modulate the local values of mean wind and temperature, and turbulence can result under certain conditions. Consider the Richardson number in terms of the background and wave values, *i. e.*,

$$Ri = (g/\Theta_r) \; [\partial(\Theta + \theta)/\partial z]/[\partial U/\partial z + \partial u/\partial z]^2 \; , \qquad (4.3.6)$$

where U and Θ are the background wind speed and potential temperature, respectively, and u and θ are the corresponding wave perturbation quantities. Gravity-wave perturbations vary in time and space more rapidly than the background quantities. Thus, at a given time or spatial location, u or θ may take on values such that $Ri \leq 1/4$ and turbulence results.

As an example of this effect, consider the case of a gravity wave launched by a uniform flow with constant stratification over a corrugated surface. The latter is described by $h(x) = H \; cos(k\,x)$, where H is the amplitude and k is the wavenumber of the corrugation. For the case $m = N/U \gg k$, where U is the velocity in the positive x-direction and m is the gravity wave's vertical wavenumber, linear wave theory gives

$$u(x, z) = -\,U\,m\,H \; sin(k\,x + m\,z) \qquad (4.3.7a)$$

and

$$\theta(x, z) = -\,H \; \partial\Theta/\partial z \; cos(k\,x + m\,z) \; . \qquad (4.3.7b)$$

Using these expressions in Eq. (4.3.6) and assuming $\partial^2\Theta/\partial z^2 \ll 1$ leads to

$$Ri = [1 + m\,H \; cos(k\,x + m\,z)]/[m\,H \; sin(k\,x + m\,z)]^2 \; . \qquad (4.3.8)$$

When $mH \geq 1$, $Ri \leq 0$ over some range of x and z and the flow will be convectively unstable there. (Note that the flow may become dynamically unstable when $Ri \leq 1/4$, and $m\,H$ need not be greater than unity for this case.) The condition that $m = N/U$ gives the requirement for turbulence as $N\,H/U \geq 1$ or $U/(N\,H) = F \leq 1$, where F is the Froude number. A critical Froude number, F_c, is seen to have a value of unity, and turbulence will occur when $F \leq F_c$.

The linear gravity waves in the SBL are described by the Taylor–Goldstein equation:

$$\partial^2\hat{w}/\partial z^2 + [\{N/(c - U)\}^2 + \{(\partial^2 U/\partial z^2)/(c - U)\} + k^2]\hat{w} = 0 \; , \qquad (4.3.9)$$

where $\hat{w}(k, z)$ is the Fourier transform of the perturbation vertical velocity, $c(k)$ is the wave phase velocity, and k is the horizontal wavenumber of the gravity wave. A vertically propagating wave may encounter a height where the local velocity equals the wave phase velocity, *i.e.*, $c = U(z_c)$.

The height z_c is called a critical level, and Eq. (4.3.9) indicates that this condition represents a singularity. Booker & Bretherton (1967) showed that, as the wave passes through the critical level, the wave stress $\tau = \rho \overline{uw}$ is reduced by an amount $exp\{-2\pi [Ri(z_c) - 0.25]^{1/2}\}$. For $Ri(z_c) = 1$, this reduction is 4.3×10^{-3}; thus, in almost all circumstances, the wave is essentially absorbed by the mean flow. When the wave is absorbed, the momentum carried by the wave decelerates the mean flow. The linear analysis of Booker & Bretherton (1967) predicts that this absorption occurs at the critical level; however, they point out that as the wave approaches z_c from below, the horizontal velocity perturbation grows as $(z_c - z)^{-1/2}$, suggesting that the wave breaks down into turbulence before it reaches the critical level. Similar asymptotic analysis shows that $Ri \propto (z_c - z)^2$, so that the wave-modulated Richardson number can fall below 0.25 some distance below z_c. Nappo & Chimonas (1992) examined the interaction of terrain generated gravity waves with critical levels located in the SBL. They showed that for typical SBL conditions and for terrain heights on the order of 10 m and higher, the wave stress on the atmosphere can be larger than the friction stress at the ground surface.

We have discussed how gravity waves can generate turbulence by creating an instability in the flow. Once produced, the turbulence is free to interact with the gravity waves and the mean flow. Analyses of these interactions has been carried out in a series of studies by Einaudi & Finnigan (1993), and references mentioned therein. The mechanism by which waves and turbulence interact and modify the mean flow is commonly referred to as "wave-turbulence coupling". Fua et al. (1982) list the four basic ideas underlying the process:

1. The occurrence of turbulence can be related to the local (wave-modified)
   Richardson number.
2. Turbulence occurs with a mean and a periodic component.
3. Turbulence extracts energy from the wave, limiting its growth, or feeds energy back into the wave.
4. Turbulence modifies the mean fields.

The analysis of wave-turbulence coupling is based on a triple decomposition of each flow variable $b(t, x_i)$ into a mean part $\bar{b}$, a periodic component $\tilde{b}$, and turbulent component b':

$$b(t, x_i) = \bar{b}(x_i) + \tilde{b}(t, x_i) + b'(t, x_i) \; . \tag{4.3.10}$$

This decomposition is useful only when it is possible to distinguish between wave and turbulence signals. We noted earlier that it may be impossible to rigorously separate waves and turbulence; however, this separation is possible if the wave can be clearly identified. Following Fua et al. (1982), a time and a phase average operator can be defined as

$$\bar{b} = \lim_{t \to \infty} \{1/(2t)\} \int_{-t}^{t} b(t') \, dt' \tag{4.3.11}$$

and

$$< b > = \bar{b} + \tilde{b} = \lim_{N \to \infty} (1/N) \sum_{n=1}^{N} b(t + n\tau) \, , \qquad (4.3.12)$$

where τ is the period of the wave. Equation (4.3.11) is the straightforward time average operator, while the second operator in Eq. (4.3.12) averages over an ensemble of points having the same phase with respect to the known gravity wave which acts as a reference oscillator.

Einaudi & Finnigan (1993) used ground-based microbarograph pressure data and linear wave theory to identify and characterize gravity waves. Phase averaging in this theory requires the gravity wave to be linear, monochromatic, long-lived (more than ten wave periods), and have constant amplitude; however, such waves in the SBL are the exception rather than the rule. This does not mean that wave-turbulence coupling events are rare in the SBL, but that the conditions for which the linear analysis is appropriate are rare. Einaudi & Finnigan (1993) summarized a number of cases of observed coupling of waves and turbulence in the SBL. They discussed the salient features of the wave-turbulence mechanism and concluded that the transfer of kinetic energy from wave to turbulence is a significant term in the TKE budget.

5 Numerical models of idealized SBL

There are no models of the SBL that can account for most of the real-world complications discussed earlier. Several first- and second-order and TKE closure models of the idealized SBL were developed during the past two decades for studies of its evolution, steady state turbulence structure, and related parameterizations. These models successfully reproduced the key boundary-layer parameters and observed turbulence characteristics under moderately stable conditions. These studies suggest that, if the surface sensible heat flux becomes invariant in time, the idealized SBL approaches a quasi-steady state within a few hours after the sunset.

5.1 First-order and TKE closure models

Businger & Arya (1974) used a simple one-dimensional model based on height and stability-dependent eddy diffusivities (§4.1.4) to solve directly for the steady-state structure. They showed that the SBL depth h followed Zilitinkevich's (1972) diagnostic equation (4.1.1), but could not obtain information on the approach to steady state. Blackadar (1976) simulated the variation of wind and temperature profiles in the SBL by including a simple model for the soil heat budget. His eddy diffusivities, specified in terms of a mixing length and the Richardson number, vanish when the latter exceeds a critical value of 0.25.

Delage (1974) used a one-dimensional numerical model based on a TKE closure to study the evolution of the SBL. The closure included a prognostic equation for the TKE, which was solved together with a diagnostic equation for the turbulence length scale:

$$\partial \bar{e}/\partial t = \partial(K_m\, \partial \bar{e}/\partial z)/\partial z + K_m\, S^2 + K_h\, N^2 - (a\,\bar{e})^{3/2}/l_m \ , \qquad (5.1.1)$$

where S and N are as defined in §4.1.3, and the eddy diffusivities K_m and K_h were derived from the relation:

$$K_{m,h} = \ell_{m,h}\,(a\,\bar{e})^{1/2} \ . \qquad (5.1.2)$$

The length scales (§4.1.3) for momentum and heat, $\ell_{m,h}$, were specified as

$$\ell_{m,h} = |\phi_{m,h}/(k\,z) + 1/\lambda|^{-1} \ , \qquad (5.1.3)$$

where $\lambda = 0.0004\,|G/f|$, and $\phi_m(z/L')$ and $\phi_h(z/L')$ were given by the M-O similarity relations (§2.2) except that the length scale L' was defined locally, analogous to Λ in Eq. (2.1.2). In this model study, horizontal advection and radiation flux divergence terms were neglected, and a realistically decreasing surface cooling rate was specified to calculate the surface temperature. Delage found that the SBL reached a steady state after several hours from evening transition when the velocity shear across the SBL decreased. Nieuwstadt & Driedonks (1979) applied this one-dimensional numerical model to simulate a case of the nocturnal boundary layer development; the model predictions compared well with the mean wind, temperature, and turbulence profiles measured along the 200 m high meteorological mast at Cabauw. Rao & Snodgrass (1981) used a similar model to investigate the evolution and structure of the nocturnal drainage flows over a simple uniformly-sloping surface.

Tjemkes & Duynkerke (1989) used a one-dimensional model similar to that described above, but including the radiation-flux divergence in the temperature equation and a rate equation for surface temperature based on Deardorff's (1978) force–restore method. The absorption and emission of infrared radiation by water vapor and carbon dioxide is simulated using a narrow band model. The calculated profiles of turbulence and mean thermodynamic variables, and surface fluxes compared satisfactorily with the Cabauw data for two cloudless nights, though the detailed structure of these profiles could not be reproduced due to local terrain inhomogeneities. The inclusion of radiative cooling within the atmosphere reduced the temperature change cross the boundary layer and increased its height by about 25 percent . Nappo & Rao (1987) developed a time-dependent two-dimensional model of the katabatic flow over a simple finite-length slope; this high-resolution numerical model, based on a TKE closure similar to that described by Eqs. (5.1.1) to (5.1.3), was used to investigate the temporal and spatial variation of the structure of katabatic flows. They showed that, for near-neutral ambient stratification, the calculated entrainment coefficient varies as a function only of the slope-Richardson number, in agreement with the laboratory data of Ellison & Turner (1959). The depth, speed, and entrainment rate of the flow decrease markedly as the ambient stratification increases.

For horizontally homogeneous SBLs, Zeman (1979) described a slab model that was derived by integrating the mean wind and temperature equation over z. This model compared favorably with a second-order closure model in simulating the evolution of the nocturnal boundary layer. The slab model is suitable for those applications where only integral properties of the boundary layer such as its depth, bulk wind and temperature, and surface turbulent fluxes are of interest, and details of the turbulence structure are not necessary.

5.2 Second-order closure models

Wyngaard (1975) and Rao & Snodgrass (1978) investigated the evolution and structure of an idealized SBL, using a full set of second-order (Reynolds stress) closure model equations and a specified SCR. The predictions of these time-dependent one-dimensional SBL models are parameterized within the framework of the surface-layer and PBL similarity theories, and successfully compared with the Minnesota (Izumi & Caughey, 1976) observations and other model results. They found that the SBL could approach steady state after 2–8 hr depending on the specified SCR, and its equilibrium depth h_e obeys Zilitinkevich's (1972) similarity prediction. Rao & Snodgrass (1979) showed that the eddy diffusivity distribution calculated from Eqs. (5.1.2) and (5.1.3) agrees well with that computed directly from the full second-order closure model. Yamada & Mellor (1975) simulated two days of Wangara ABL data (Clarke *et al.*, 1971) using a Level 3 model which included predictive equations only for TKE, and variances of temperature and water vapor mixing ratio; the remaining turbulence moments were calculated from simpler diagnostic equations. The computed wind and virtual potential temperature agreed with the observed values.

Brost & Wyngaard (1978) developed a simplified "diagnostic" version of Wyngaard's (1975) model by neglecting the time change, Coriolis, and turbulent transport terms, and by using only some of the full set of equations. The model was closed in terms of a turbulent integral length scale defined in Eq. (4.1.11), and was shown to reproduce the M-O similarity in the surface layer and give monotonically decreasing turbulence profiles and monotonically increasing wind and temperature profiles above the surface layer. Brost & Wyngaard found that the SBL does not reach steady state when the lower boundary condition on temperature is calculated from a surface energy budget, because of the slowness of the SBL in adjusting to the changing surface conditions. They also found that the terrain slope has a strong effect on SBL depth depending on the wind direction. Caughey *et al.* (1979) showed that the dissipation rates of TKE and temperature variance, and the SBL depth predicted by the Brost-Wyngaard model agreed well with observed values in the 1973 Minnesota experiment.

Nieuwstadt (1984) used parameterized equations for the turbulent variances and covariances similar to those of Brost and Wyngaard (1978), and developed a SBL theory based on local scaling hypothesis (§2.1). The latter permits dimensionless scaled variables to be expressed as functions of a single parameter, z/Λ, where Λ is the local similarity length scale given by Eq. (2.1.2). In the limit of $z/\Lambda \to \infty$, the locally-scaled dimensionless variables approach a constant value (z-less stratification), and both the gradient and flux Richardson numbers become constant in the SBL. Nieuwstadt showed that the local scaling hypothesis was supported by the Cabauw observations. Derbyshire (1990) extended this quasi-steady model and found a maximum value of the surface heat flux which depends only on the synoptic parameters, and suggested that radiative cooling above this threshold value would cause intermittency.

Rodi (1985) and Duynkerke (1988), among others, described simpler $\bar{e} - \epsilon$ turbulence closure models applicable to the SBL. Rodi tested his model with laboratory data for a number of horizontal shear layers with stable stratification. The reduction of mixing and spreading of the shear layer due to stratification was well simulated by the model, using the same model constants in all cases. Duynkerke showed that his model can reproduce the second-order closure model results of Wyngaard (1975) and Brost & Wyngaard (1978) for a constant SCR.

5.3 Large eddy simulation

Unlike in the convective boundary layer, large eddy simulation (LES) has not been widely used in the SBL since much of the turbulence resides in the subgrid scales, and its parameterization becomes critically important. In addition, LES does not work for strongly stratified conditions with patchy turbulence. Because of the smaller characteristic large-eddy length scale, computers were not powerful enough until recently to permit simulation of the turbulence in the SBL. However, LES can become a most promising tool in future research on SBL structure, especially in providing "data" free from mesoscale effects and in developing suitable parameterizations for simpler models.

Mason & Derbyshire (1990) used a Smagorinsky-type subgrid model to simulate the SBL with 40 x 32 x 62 mesh points in a domain of 500 x 300 x 1000 m in x, y, z directions, respectively. A filter, conceptually separate from the numerical mesh spacing, was used to distinguish resolved and subgrid motions. Mason & Derbyshire discussed the technical requirements for SBL simulations and the characteristics of the simulated SBL. Results showed the development of a shear-driven SBL, with little sign of distinctively wave-like motions. The flow statistics were consistent with local scaling, and that framework was used to compare with other data and theoretical models.

In boundary layer flows, LES can show significant departures from the observed profiles in the near-surface region. Mason & Thomson (1992) showed that the inclusion of stochastic variability in the model for subgrid-scale (SGS) stress leads to a marked improvement in the near-wall flow simulation. Brown et al. (1994) repeated and extended Mason & Derbyshire's (1990) LES work for SBLs, using an improved stochastic SGS model parameterization over a wider range of stabilities. They found more turbulence and a deeper boundary layer, due to the tendency of the new parameterization to increase the surface stress. Andren (1995) described the LES study of a weakly stratified SBL using a different SGS modeling approach, which he showed is about as successful in improving the near-surface flow behavior as the computationally more expensive approach of including random SGS stresses. It was found that bursts dominate the vertical turbulent fluxes.

5.4 Full turbulence simulation

Full turbulence simulation (FTS) solves the Navier-Stokes equations directly without any averaging or using any models. The number of grid points required to resolve all the turbulent eddies in a flow is of the order $\sim (\ell/\eta)^3$, where ℓ/η is the ratio of the largest to the smallest turbulence scales. This number is too large for atmospheric turbulent flows, and is beyond the capability of current-generation supercomputers. Since the increase in computing requirements is proportional to $(Re)^3$, FTS has been used so far only for very low Reynolds number ($Re \sim 10^2$) flows, under the assumption that statistics of large scales are relatively insensitive to Re. Coleman et al. (1992) described such an FTS study of the stably stratified turbulent Ekman layer. They found that some of the simulation data, when nondimensionalized according to the local scaling scheme (§2.1), agreed well with the atmospheric measurements, and there was good agreement between the FTS results and Mason & Derbyshire's (1990) LES results.

6 Atmospheric Diffusion

Atmospheric diffusion under stable conditions is an important topic for both research and applications. Work in this area has been generally limited to conditions most amenable to observations and theoretical analysis, such as near-surface releases of tracer materials and surface-layer similarity theory. It is not clear that the latter holds under conditions of light winds and strong static stability, and yet these are the conditions that lead to very high air pollution concentrations. Winds can be highly variable in time and location at night, and this variability makes predictions of transport and diffusion from elevated releases difficult when observations are not available. A discussion of the fundamentals and details of atmospheric diffusion and its modeling is not possible here; however, these can be found in several texts including Nieuwstadt & van Dop (1982), Hanna et al. (1982), Pasquill & Smith (1983), Panofsky & Dutton (1984), and Venkatram & Wyngaard (1988).

6.1 Diffusion models

We limit the discussion here to two general types of diffusion models: Gaussian plume models, and Lagrangian particle diffusion models.

6.1.1 Gaussian plume models

The semi-empirical plume or puff diffusion models are widely used in air pollution forecasting and regulation. These models assume a specific distribution of concentration within the pollutant cloud. The Gaussian (normal) is the most commonly used distribution. The Gaussian plume formula for a continuous point source is given by

$$C(x, y, z) = [Q/(2\pi \sigma_y \sigma_z U)] \, e^{(-y^2/2\sigma_y^2)} \, \{e^{-(z-H)^2/2\sigma_z^2} + e^{-(z+H)^2/2\sigma_z^2}\} \,, \quad (6.1.1)$$

where C is the concentration (mass per unit volume of air), Q is the source strength (mass per unit time), U is the mean wind speed, H is the effective height of the plume, and (x, y, z) are coordinates of the receptor location. The x-axis is directed along the mean wind vector, and y-axis is taken normal to the mean wind in the horizontal plane. The dispersion parameters σ_y and σ_z correspond to the standard deviations of the concentration distributions in the y and z directions, respectively. The last term in Eq. (6.1.1) accounts for the lower boundary by assuming perfect reflection of diffusing material at the ground surface.

Turbulence is implicitly contained in the dispersion parameters σ_y and σ_z, which are complex functions of static stability, wind speed, and travel distance, and are based on combinations of experimental results and theory (see, e.g., Hanna et al., 1982; Pasquill & Smith, 1983). Taylor (1921) derived a diffusion formula for homogeneous and stationary turbulence and neutral static stability:

$$\sigma_y^2 = 2\sigma_v^2 \int_0^t (t - t') \, R_y(t') dt' \,, \quad (6.1.2)$$

where σ_v is the rms turbulence velocity in the y direction, t is the travel time (x/U), and $R_y(t') = \overline{v(t)v(t+t')}/\sigma_v^2$ is the Lagrangian autocorrelation coefficient. For time $t' \ll \tau_i$, where τ_i is the integral time scale, $R_y \to 1$; for $t' \gg \tau_i$, $R_y \to 0$. Thus, without any assumptions regarding $R_y(t')$, two simple deductions follow:

$$\sigma_y^2 = \sigma_v^2 t^2 \,, \quad (6.1.3)$$

for small t, and

$$\sigma_y^2 = 2\sigma_v^2 \tau_i t \quad (6.1.4)$$

for large t. The last relation results from using Eq. (4.3.2) after noting that $t'/t \ll 1$.

Under stable conditions, σ_v can be quite sensitive to local topography, mesoscale turbulence, and averaging times, and the use of similarity theory to evaluate σ_v can be uncertain. If surface layer measurements of σ_v are available, then σ_y can be evaluated from empirical relations such as

$$\sigma_y = \sigma_v t/(1 + x/x_o)^{1/2} \,. \quad (6.1.5)$$

Briggs (1973) gave $x_o = 10^4$ m, but cautioned that the empirical fit was good only to $x = 10^4$ m. The lack of empirical support for a $t^{1/2}$ or $x^{1/2}$

asymptote at large x for σ_y in the atmosphere is thought to result from there being no upper limit to horizontal eddy sizes, short of hemispheric scales. If the rms horizontal wind direction fluctuations (σ_θ) are small, then $\sigma_v \approx \sigma_\theta U_a$ where U_a is the average wind speed in the surface layer. Hanna (1990) noted that σ_v is of the order of 0.5 m s^{-1} at night at a wide variety of sites. Based on this, Briggs (1996) suggested that $\sigma_y = 0.5\,x/U$ provides a robust estimate in the absence of site-specific data.

Using an equation analogous to (6.1.4), we can write

$$d\sigma_z^2/dt = 2\,K_z \ , \qquad (6.1.6)$$

where $K_z = \sigma_w^2\,\tau_i$ is the eddy diffusivity for plume material. Using $K_z = K_m = k\,u_*\,z/\phi_m$ from surface-layer similarity theory (§4.1.4), we get

$$\sigma_z\,d\sigma_z/dt = k\,u_*\,z\,(1 + \beta_m\,z/L)^{-1} \ . \qquad (6.1.7)$$

In the surface layer, the vertical spread of a plume will be controlled by eddies of about the same size as the plume. For dispersion from a surface source, we can approximate σ_z with z (Venkatram, 1988) and write Eq. (6.1.7) as

$$d\sigma_z/dt = k\,u_*\,(1 + \beta_m\,\sigma_z/L)^{-1} \ . \qquad (6.1.8)$$

For $\sigma_z \ll L$, this leads to

$$\sigma_z \sim u_*\,t \sim u_*\,x/U_e \ , \qquad (6.1.9)$$

where U_e is an effective velocity in the surface layer. For $\sigma_z \gg L$, the effects of static stability limit the size of eddies to L. This makes the diffusion quasi-Fickian and leads to

$$\sigma_z^2 \sim u_*\,L\,t \sim u_*\,L\,x/U_t \ , \qquad (6.1.10)$$

where U_t is a transport velocity above the surface layer. From similarity theory, for $z \gg L$, $U(z) \sim u_*z/L$; this implies $U_t \sim u_*\sigma_z/L$. Substituting this in Eq. (6.1.10), we get

$$\sigma_z \sim L^{2/3}\,x^{1/3} \ . \qquad (6.1.11)$$

We see that, for short travel times, Eq. (6.1.9) is in agreement with Taylor's diffusion theory, Eq. (6.1.3), i.e., $\sigma_z \sim t$; however, for long travel times (i.e., $\sigma_z \gg L$), similarity theory predicts a $t^{1/3}$ growth rate while Taylor's theory predicts a $t^{1/2}$ growth rate from Eq. (6.1.10). The slower rate of growth given by Eq. (6.1.11) results from the smaller vertical spread in the strong shear layer for $z \gg L$.

The modeling of diffusion in the upper SBL is complicated by a lack of direct meteorological measurements and diffusion experiments. When the upper SBL is disconnected from the surface layer, the surface-layer similarity theory no longer applies. Although a comprehensive understanding of diffusion in stably-stratified turbulence is yet to be attained, progress has been made in our ability to model plume diffusion. The horizontal spread of a plume is directly affected by static stability, but the mechanisms are not yet understood. Instead, use is often made of empirical relations (Draxler, 1976) such as

$$\sigma_y = \sigma_v t/[1 + 0.9 (t/T_i)] , \qquad (6.1.12)$$

where $T_i = 1000$ s for elevated sources in a stable environment. Alternately, Phillips & Panofsky (1982) proposed:

$$\sigma_y = 0.617 \sigma_v t [(T_i/t) - \{(T_i/t)^2/5.25\} \, ln(1 + 5.25 t/T_i)]^{1/2} , \qquad (6.1.13)$$

where $T_i = 5.25 \tau_y$, and τ_y is the Lagrangian integral time scale for horizontal turbulence. These types of relations are based on analyses of diffusion experiments, and consequently they may be applicable only to specific terrains and meteorological conditions. Also, it may not always be possible to specify the required values of turbulence parameters such as σ_v or σ_θ, and integral time scales, etc. However, with the increasing development of remote sensing systems, $e.g.$, Doppler radar and lidar, these quantities can be now measured directly.

The vertical spread of plumes in the upper SBL will be strongly influenced by the stable stratification. Pearson et $al.$ (1983) used a theoretical analysis of stable shear-free flow to show that:
  1. buoyant forces restrict the direct eddy motions to vertical distances on the order of σ_w/N;
  2. continued vertical spread occurs through diffusive exchange of density between fluid parcels;
  3. stable stratification and possibly the presence of gravity waves can cause the autocorrelation for vertical velocities to have negative values, and this will lead to small or even zero values of the integral time scale τ_i. When this happens, σ_z will grow at a rate less than $t^{1/2}$ or may even be constant.

Using these ideas, Hunt (1982) proposed that

$$\sigma_z = \sigma_w t \qquad (6.1.14a)$$

for short travel times, and

$$\sigma_z = (\sigma_w/N) (\zeta_z^2 + 2\gamma^2 N t)^{1/2} \qquad (6.1.14b)$$

for long travel times, where ζ_z is a constant equal to about 1.3, and γ is a mixing coefficient. Weil (1988) gave an interpolation formula between these two regimes as

$$\sigma_z = \sigma_w t \, [1 + N^2 t^2/(1 + 2\gamma^2 N t)]^{-1/2} . \qquad (6.1.15)$$

Equation (6.1.15) implies linear growth at short travel times when γ has little effect. At larger times, however, the plume growth is governed by the value of γ. If γ is large (indicating rapid mixing of plume material with ambient air), then plume growth is parabolic. If γ is small, then plume growth will be limited by buoyancy effects; plume will grow slowly, but will eventually reach the parabolic growth rate. If $\gamma = 0$, then $\sigma_z = \sigma_w/N$ at large t, and the plume thickness is constant. Thus, γ is an important parameter in SBL diffusion. Values of γ have been estimated from turbulent heat flux measurements, and these range between 0.1 and 0.4.

In contrast to Weil's (1988) formula, Venkatram (1988) developed a formula for σ_z which does not include fluid mixing, but interpolates between linear and parabolic growth rates, and uses a theoretical expression for the Lagrangian time scale:

$$\sigma_z = \sigma_w t \left[1 + t/(2\tau_i) \right]^{-1/2} , \tag{6.1.16}$$

where

$$\tau_i = l/\sigma_w . \tag{6.1.17}$$

Following Delage (1974), Brost & Wyngaard (1978), and others (§4.1.3), an equation for the length l in Eq. (6.1.17) is obtained by interpolating between the neutral (l_n) and very stable (l_s) limits, i.e.,

$$1/l = 1/l_n + 1/l_s , \tag{6.1.18}$$

where $l_n = \alpha_n z_r$, $l_s = \gamma_s^2 \sigma_w/N$, z_r is the release (source) height, and α_n and γ_s are constants. Venkatram used theoretical arguments to determine the values $\alpha_n = 0.36$ and $\gamma_s = 0.52$.

6.1.2 Lagrangian particle diffusion models

Semi-empirical diffusion models such as those described in the previous section ultimately require experiments to fix the dispersion parameter values. However, even when these data are available, the calculated values may only be applicable to specific locations and atmospheric conditions. An attractive alternative approach is provided by Lagrangian particle diffusion models (also referred to as stochastic, Langevin, Monte Carlo, and random-walk models). Sawford (1993) gave an overview of some recent developments in this modeling technique. These models replace a continuous distribution of pollutant mass with a large number of advecting mass points or fluid particles. Given the mean and statistical properties of a turbulent flow, the trajectory of each particle is calculated by assuming its total motion to be the sum of deterministic and random components. A mathematical representation is given by

$$x_i(t + \Delta t) = x_i(t) + [U_i(\vec{x}, t) + a_i(\vec{x}, \vec{U}, t) + b_{ij}(\vec{x}, \vec{U}, t) \, d\omega] \, \Delta t , \tag{6.1.20}$$

where $\vec{U}$ is the mean wind velocity vector, and a_i and b_{ij} are the drift (deterministic) and diffusion (random) terms, respectively; $d\omega$ is a component of a Gaussian white noise probability distribution which is uncorrelated with the other components and is uncorrelated in time, and Δt is the time step for each trajectory. The problem is to determine the functions a and b for a particular turbulent flow field for which the Eulerian flow statistics are given. Examples of these functions are given in Wilson & Sawford (1996). These trajectories correspond to different flow realizations. The size of the particles is assumed to be large compared to the average distance between fluid molecules, but small enough to follow the eddies of a turbulent flow without being deformed. Particle trajectories are calculated in an Eulerian grid with cell dimensions $\Delta x, \Delta y$, and Δz. The average concentration in a given cell during a sampling period ΔT is given (Luhar & Rao, 1994) by

$$C(x,y,z,t) \,=\, \lfloor q_o/(\Delta x \, \Delta y \, \Delta z \, \Delta T) \rfloor \, \Sigma_{j=1}^{n_p} t_j \,, \qquad (6.1.21)$$

where q_o is the mass assigned to a single particle, n_p is the number of particles counted in the cell during the period ΔT, and t_j is the total time spent by the jth particle in the cell during this period.

Luhar & Rao (1993) used a Lagrangian stochastic model to examine dispersion in a katabatic flow over a simple (uniform) slope, and in the nocturnal drainage flow over a complex mountain slope. Three models for the Lagrangian velocities were tested. In Model 1, it is assumed that turbulent velocities of the particle are uncorrelated in time, and therefore can be modeled by a white-noise process, $i.e.$, a pure random-walk model. The height $Z(t)$ of a particle in an inhomogeneous Gaussian turbulent flow is then given (Durbin, 1983) by

$$dZ \,=\, \{w + \partial K_z(z)/\partial z\} \, dt \,+\, \sqrt{2K_z(z)} \, d\xi \,, \qquad (6.1.22a)$$

where w is the mean Eulerian vertical velocity (which need not be zero in horizontally inhomogeneous flows), and K_z is the vertical eddy-diffusivity; here, following Luhar & Rao (1993), we use capital letters to denote Lagrangian quantities and lower case letters to denote Eulerian quantities. The quantity $d\xi(t)$, with units of $(\text{time})^{1/2}$, is a Gaussian random forcing with a mean of zero, a variance dt, and a covariance of zero between subsequent random events. If the streamwise diffusion is small compared to transport, then the increment in the particle's downwind direction is

$$dX \,=\, u \, dt \,, \qquad (6.1.22b)$$

where u is the mean Eulerian velocity. In a three-dimensional problem, a similar equation can be written for the y-component.

There are many atmospheric turbulent flows in which the Lagrangian time scale τ is very large ($e.g.$, τ is about 10 min in a convective boundary layer), and then the velocity correlation must be considered. In Model 2, it is assumed that the particle velocities are correlated up to a time

τ, but the particle accelerations are uncorrelated in time and can be modeled by a white-noise process. The assumption that the accelerations are uncorrelated is a good one in high Reynolds number flows, because the correlation time scale of particle accelerations is on the order of the Kolmogorov time scale, $\tau_\eta = (Re)^{-1/2}\tau$. Then, particle accelerations can be considered uncorrelated for time steps that lie in the inertial subrange, i.e., $\tau_\eta \ll dt \ll \tau$. The fluctuating particle velocity in the vertical direction is then given (Thomson, 1987; Luhar & Britter, 1989) by

$$dW = 0.5\left\{-C_o\,\epsilon\,W/\sigma_w^2 + (1 + W^2/\sigma_w^2)\,\partial\sigma_w{}^2/\partial z\right\}dt + \sqrt{C_o\,\epsilon}\,d\xi(t)\ ,\quad (6.1.23a)$$

where $C_o = 2\sigma_w{}^2/(\epsilon\tau)$ is a constant. Again assuming that the streamwise diffusion is small compared to the mean transport, the incremental particle displacements are calculated using

$$dZ = (W + w)\,dt \qquad\qquad (6.1.23b)$$

and

$$dX = u\,dt\ . \qquad\qquad (6.1.23c)$$

Model 3 is the same as Model 2, but includes the streamwise diffusion. An equation similar to (6.1.23a) is used to calculate dU, after replacing W with U and σ_w with σ_u. The downwind particle displacement is then given by

$$dX = (U + u)\,dt\ . \qquad\qquad (6.1.23d)$$

Luhar & Rao (1993) tested the three models with a simulation of dispersion from a continuous line source in a 2-D katabatic flow along a simple slope. They used Eulerian velocities and turbulence statistics derived from a 2-D katabatic flow model developed by Nappo & Rao (1987). All three particle models gave similar values for the dispersion parameters, and agreed well with the results of a 2-D Eulerian dispersion model developed by Nappo et al. (1989). Luhar & Rao (1994) applied Model 1 to simulate the ground-level concentrations from tracer releases in 3-D drainage flows, using measured Eulerian velocities and turbulence statistics; their results agreed quite well with the observed downwind concentrations. These results indicate that Lagrangian particle dispersion models give realistic simulations of atmospheric diffusion, and are potentially very useful in nonhomogeneous flows.

Lagrangian dispersion models require Eulerian velocities and turbulence statistics; however, these are generally not directly available, and must be inferred using similarity theory or other scaling laws. Recently, Kemp & Thomson (1996) have combined the LES model (§5.3) developed by Brown et al. (1994) with a Lagrangian particle diffusion model to calculate atmospheric dispersion parameters in a SBL flow. Random particle displacements in each of the three coordinate directions were calculated using a Gaussian distribution of displacements with a standard deviation σ_d given by

$$\sigma_d{}^2 = 2\,K_i\,\delta t\ , \qquad\qquad (6.1.24)$$

where K_i is the interpolated eddy diffusivity used in the LES model. This approach corresponds to Model 1 of Luhar & Rao (1993). Agreement between the numerical results and the dispersion theory of Pearson et al. (1983), Eq. (6.1.14), was reasonable in terms of magnitude, particularly at short times. However, the LES did not reproduce the expected vertical plume growth behavior, namely, a noticeable inflection in the σ_z versus t curve after travel time $t \sim 1/N$ anticipated from the theory. Kemp & Thomson (1996) have suggested that the LES cannot truly represent the motions of individual particles well enough to reproduce this theoretical behavior.

6.2 Stagnation Diffusion

Stagnation occurs when the synoptic pressure gradients are weak enough to have no measurable effect on the boundary layer wind. Under stagnation conditions, the boundary-layer flow and turbulence will be driven mostly by the local terrain and the surface heat-flux induced buoyancy (Briggs, 1992). In the presence of strong thermal stability, these conditions present the"worst case" scenario for public safety, yet conventional diffusion theory and similarity scaling generally fail under these conditions. The Bhopal gas tragedy (Rao et al., 1986; Singh & Ghosh, 1987; Sharan et al., 1995) occurred during light winds and strong stability, resulting in over 2500 deaths. For very light winds, the partitioning of the diffusion process into transport and dispersion becomes difficult. This is because the concept of a mean wind, which implies the existence of an organized flow over a spatial domain, is not valid under these conditions. An average wind calculated at a given location may be meaningful only at that location. Most of diffusion theory is based on Taylor's frozen turbulence hypothesis, which holds only when the turbulence intensity σ_u/U is small; this is generally not the case in light winds. Similarity theories used in diffusion modeling assume continuous turbulence with a well-defined inertial subrange. However, these assumptions are seldom realized during stagnation conditions. Large low-frequency oscillations in the horizontal wind direction are often seen in the SBL, especially when the winds are light. These oscillations not only effect plume transport, but also the calculation of dispersion parameters used in Gaussian plume models. In this section, we will briefly discuss these problem areas.

Statistical models of diffusion neglect along-wind diffusion in comparison to advection by the mean wind. This assumption is valid for moderate winds ($U > 2$ m s^{-1}) and moderate turbulence ($\sigma_u/U \leq 0.3$). However, in weak winds and strong turbulence ($\sigma_u/U \geq 0.3$), the longitudinal diffusion becomes important especially near the source. Also, the estimation of plume spread based on Gaussian dispersion parameters is intended to be applied specifically to moderate winds ($U > 2$ m s^{-1}); as a result, Gaussian models generally produce an unreasonable overestimation of concentrations in low wind conditions (see, e.g., Zannetti, 1986). For this reason, it is recommended that gradient-transport models be used (Arya, 1995; Sharan et al., 1996). However, the use of constant eddy diffusivities is not appropriate under weak winds, since the diffusivities are based on the existence of large-eddy length and velocity scales. Arya (1995) suggested that eddy diffusivities should be considered as functions not only of turbulence, but also of distance from the source. He developed interpolation formulas for eddy diffusivities which are consistent for short and long downwind distances:

$$K_x = \sigma_u^2 \, (x/U) \, [1 + x/(c \, L_x)]^{-1} \, . \tag{6.1.25}$$

Here, c is a constant of order 1 and $L_x = U \tau_u$, where τ_u is the Lagrangian time scale for velocity fluctuations in the x-direction. Similar expressions can be written for K_y and K_z.

Among characteristic features of the SBL during light variable wind conditions are the large-amplitude low-frequency fluctuations of horizontal wind direction (*e.g.*, Panofsky *et al.*, 1978; Hanna, 1983; Rao & Schaub, 1990; Etling, 1990; Leahey *et al.*, 1994). These mesoscale wind oscillations often cause meandering of pollutant plumes downwind of elevated point sources. These meanders result in relatively high lateral turbulence intensities and diffusion rates when averaged over an hour. Field experiments and theoretical considerations (Kristensen *et al.*, 1981; Hanna, 1983) show that averaged plume concentrations under meandering conditions can be 2 to 6 times lower than for straight-line plumes. Oscillation periods of these fluctuations have been observed to range from 5 min to several hours. The exact cause of low frequency wind direction fluctuations is not yet clear, and it is quite likely that several different mechanisms are responsible (Etling, 1990). For example, over flat terrain or coastal regions, it is believed that gravity waves or vortices with horizontal or vertical axis are the cause (Sethuraman, 1980; Raynor & Hayes, 1984); over complex terrain, meandering may be due to terrain-induced vortices or temporal variations in drainage flows (Hanna, 1983; Leahey *et al.*, 1988). However, meandering is not always observed for given meteorological conditions, and this suggests that it is not entirely due to local boundary-layer processes.

7 Conclusions

The vast literature and wide range of applications of the stable boundary layer studies in many disciplines of fluid mechanics make it difficult to provide a comprehensive description. In this chapter, we discussed selected features of the idealized stable atmospheric boundary layers using similarity theories, observations from field experiments, and numerical model results. We summarized the complicating factors in the real-world SBL, and reviewed some of the recent research results and advances in numerical simulation.

Although considerable research has been done on turbulence and diffusion in the SBL, it is fair to say that a comprehensive understanding of these phenomena has not yet been achieved. Unlike the convective boundary layer, the SBL is not easily characterized by time and space scales. Because of this difficulty, several definitions exist for the depth of the SBL. At night, above a relatively shallow surface friction layer, turbulence is observed to be sporadic and intermittent, and often exists simultaneously with gravity waves. It is known that breakdowns of the SBL often occur, resulting in bursts of heat, momentum, and pollutant-mixing near the ground surface. Numerical models of the SBL are unsatisfactory for real-world situations, because of uncertainty in the parameterizations of surface-layer turbulent fluxes. Measurements of diffusion in the SBL, especially under low wind speed conditions, often lead to highly scattered results. Predictions of dispersion under these conditions are often quite inaccurate; this is most likely due to a lack of understanding of the turbulence structure.

Field observations in the SBL are often of limited spatial and temporal resolution, whereas experience indicates that even in mildly nonuniform terrain, the SBL flow can be horizontally variable. Thus, measurements made at a single location may not be representative of the other locations or of the area-averaged data. Analyses of field data are further complicated by a lack of precise methods of identifying periods of sporadic and intermittent turbulence. Techniques such as spectral analysis, band-pass filtering, principal component analysis, and wavelet transforms can be used, but it is not clear that any one of these methods is better than any other.

On the other hand, considerable research has been performed or is currently in progress in the nocturnal boundary layers over complex terrain, wave-turbulence interactions, and other complicating features. This work includes both field and laboratory experiments. There are many advances in remote sensing and other instrumentation. Most field programs now routinely measure second and higher moments of turbulence. With the rapidly increasing power and speed of computers, large-eddy simulations as well as full-turbulence simulations of the SBL are also being taken up for simple situations. Thus, it seems reasonable to expect that our knowledge of the SBL behavior and turbulence structure, as well as our ability to simulate them, will continue to improve steadily in the future.

Acknowledgements

The support of the Air Resources Laboratory of National Oceanic and Atmospheric Administration (NOAA) during the preparation of this article is gratefully acknowledged. We are indebted to Dr. Gary Briggs of NOAA's Atmospheric Sciences Modeling Division at the U.S. EPA, and to Dr. Larry Mahrt of the Department of Atmospheric Sciences, Oregon State University, for their critical review of the draft manuscript and helpful suggestions to improve it. Several figures in this chapter are reproduced, as indicated in the captions, with kind permission from their authors and Kluwer Academic Publishers, Dordrecht, The Netherlands.

References

André, J.C. & Mahrt, L., 1982: The nocturnal surface inversion and influence of clear-air radiative cooling. *J. Atmos. Sci.* **39**, 864–878.

André, J.C., Pettré, P., Wendler, G. & Zephoris, M., 1993: Vertical structure and downslope evolution of Antarctic katabatic flows. *Waves and Turbulence in Stably Stratified Flows*, S.D. Mobbs & J.C. King, Eds., Clarendon Press, Oxford, 91–104.

Andren, A., 1995: The structure of stably stratified atmospheric boundary layers: A large-eddy simulation study. *Quart. J. Roy. Meteor. Soc.* **121**, 961–985.

Arya, S.P.S., 1975: Geostrophic drag and heat transfer relations for the atmospheric boundary layer. *Quart. J. Roy. Meteor. Soc.* **101**, 147–161.

Arya, S.P.S., 1981: Parameterizing the height of the stable atmospheric boundary layer. *J. Appl. Meteor.* **20**, 1192–1202.

Arya, S.P.S., 1988: *Introduction to Micrometeorology*. Academic Press, New York, 303 pp.

Arya, S.P.S., 1995: Modeling and parameterization of near-source diffusion in weak winds. *J. Appl. Meteor.* **34**, 1112–1122.

Blackadar, A.K., 1957: Boundary layer wind maxima and their significance for the growth of the nocturnal inversion. *Bull. Amer. Meteor. Soc.* **38**, 283–290.

Blackadar, A.K., 1976: Modeling the nocturnal boundary layer. 3rd Symposium on Atmospheric Diffusion and Air Quality, Raleigh. Preprints, Amer. Meteor. Soc., Boston, MA, 46–49.

Booker, J.R. & Bretherton, F.P., 1967: The critical level for internal gravity waves in a shear flow. *J. Fluid Mech.* **27**, 513–539.

Bretherton, F.P., 1969: Waves and turbulence in stably stratified fluids. *Radio Science* **4**, 1279–1287.

Briggs, G.A., 1973: Diffusion estimations for small emissions. 1973 Annual Report ATDL-106, Air Resources Atmos. Turb. and Diffusion Lab., NOAA, Oak Ridge, TN.

Briggs, G.A., 1992: Stagnation diffusion observed in a deeply pooling valley during STAGMAP. 10th Symposium on Turbulence and Diffusion, Portland. Preprints, Amer. Meteor. Soc., Boston, MA, 155–162.

Briggs, G.A., 1996: Personal communication.

Brost, R.A.& Wyngaard, J.C., 1978: A model study of the stably stratified planetary boundary layer. *J. Atmos. Sci.* **35**, 1427–1440.

Brown, A.R., Derbyshire, S.H. & Mason, P.J., 1994: Large eddy simulation of stable atmospheric boundary layers with a revised stochastic subgrid model. *Quart. J. Roy. Meteor. Soc.* **120**, 1485–1512.

Businger, J.A., Wyngaard, J.C., Izumi, Y. & Bradley, E.F., 1971: Flux-profile relationships in the atmospheric surface layer, *J. Atmos. Sci.* **28**, 181–189.

Businger, J.A., 1973: Turbulent transfer in the atmospheric surface layer. *Workshop on Micrometeorology*, D.A. Haugen, Ed., Amer. Meteor. Soc., Boston, MA, 67–100.

Businger, J.A. & Arya, S.P.S., 1974: The height of the mixed layer in the stably stratified planetary boundary layer. *Adv. Geophys.* **18A**, Academic Press, New York, 73–92.

Caughey, S.J., Wyngaard, J.C. & Kaimal, J.C., 1979: Turbulence in the evolving stable boundary layer. *J. Atmos. Sci.* **36**, 1041–1052.

Caughey, S.J., 1982: Observed characteristics of the atmospheric boundary layer. *Atmospheric Turbulence and Air Pollution Modelling*, F. Nieuwstadt & H. van Dop, Eds., D. Reidel Publishing, Boston, MA, 107–158.

Chimonas, G., 1972: The stability of a coupled wave-turbulence system in a parallel shear flow. *Bound.-Layer Meteor.* **2**, 444–452.

Chimonas, G. & Nappo, C.J., 1989: Wave drag in the planetary boundary layer over complex terrain. *Bound.-Layer Meteor.* **47**, 217–232.

Clarke, R.H., Dyer, A.J., Brook, R.R., Reid, D.G. & Troup, A.J., 1971: The Wangara experiment: Boundary layer data. Tech. Paper No. 19, Div. of Meteorol. Phys., CSIRO, Melbourne, 350 pp.

Coleman, G.N, Ferziger, J.H. & Spalart, P.R., 1992: Direct simulation of the stably stratified turbulent Ekman layer. *J. Fluid Mech.* **244**, 677–712.

Deardorff, J.W., 1978: Efficient prediction of ground surface temperature and moisture, with inclusion of a layer of vegetation. *J. Geophys. Res.* **83**, 1889–1903.

Delage, Y., 1974: A numerical study of the nocturnal atmospheric boundary layer. *Quart. J. Roy. Meteor. Soc.* **100**, 351–364.

Derbyshire, S., 1990: Nieuwstadt's stable boundary layer revisited. *Quart. J. Roy. Meteor. Soc.* **116**, 127–158.

Derbyshire, S., 1995: Stable boundary layers: observations, models and variability Part I: Modelling and measurements. *Bound.-Layer Meteor.* **74**, 19–54.

Draxler, R.R., 1976: Determination of atmospheric diffusion parameters. *Atmos. Environ.* **100**, 99–105.

Durbin, P.A., 1983: Stochastic differential equations and turbulent dispersion. Ref. Pub. No. 1103, Scientific and Technical Information Branch, NASA, Washington, DC.

Duynkerke, P.G., 1988: Application of the $\bar{e} - \epsilon$ turbulence closure model to the neutral and stable boundary layer. *J. Atmos. Sci.* **45**, 865–880.

Einaudi, F., Bedard, Jr., A.J. & Finnigan J.J., 1989: A climatology of gravity waves and other coherent disturbances at the Boulder atmospheric observatory during March-April, 1984. *J. Atmos. Sci.* **46**, 303–329.

Einaudi, F. & Finnigan, J.J., 1993: Wave-turbulence dynamics in the stably stratified boundary layer. *J. Atmos. Sci.* **50**, 1841–1864.

Ellison, T.H. & Turner, J.S., 1959: Turbulent entrainment in stratified flows. *J. Fluid Mech.* **6**, 423–448.

Estournel, C. & Guedalia, D., 1985: Influence of geostrophic wind on atmospheric nocturnal cooling. *J. Atmos. Sci.* **42**, 2695–2698.

Estournel, C., Vehil, R. & Guedalia, D., 1986: An observational study of radiative and turbulent cooling in the nocturnal boundary layer. *Bound.-Layer Meteor.* **34**, 55–62.

Etling, D., 1990: On plume meandering under stable conditions. *Atmos. Environ.* **24A**, 1979–1985.

Finnigan, J.J., Einaudi, F. & Fua, D., 1984: The interaction between an internal gravity wave and turbulence in the stably-stratified nocturnal boundary layer. *J. Atmos. Sci.* **41**, 2409–2436.

Finnigan, J.J., 1988: Kinetic energy transfer between internal gravity waves and turbulence. *J. Atmos. Sci.* **45**, 486–505.

Fua, D., Chimonas, G. & Zeman, O., 1982: An analysis of wave-turbulence interaction. *J. Atmos. Sci.* **39**, 2450–2463.

Garratt, J.R. & Brost, R.A., 1982: Radiative cooling effects within and above the nocturnal boundary layer. *J. Atmos. Sci.* **38**, 2730–2746.

Garratt, J.R., 1992: *The Atmospheric Boundary Layer.* Cambridge University Press, New York, 316 pp.

Garratt, J.R. & Hicks, B.B., 1990: Micrometeorological and PBL experiments in Australia. *Bound.-Layer Meteor.* **50**, 11–29.

Hanna, S.R., Briggs, G.A. & Hosker, Jr., R.P., 1982: *Handbook on Atmospheric Diffusion.* DOE/TIC-11223, NTIS, U.S. Dept. of Commerce, Springfield, VA, 102 pp.

Hanna, S.R., 1983: Lateral turbulence intensity and plume meandering during stable conditions. *J. Clim. Appl. Meteor.* **22**, 1424–1430.

Hanna, S.R., 1990: Lateral dispersion in light-wind stable conditions. *Il Nuovo Cimento* **13**, 889–894.

Haugen, D.A., 1973: *Workshop on Micrometeorology*, Ed. Amer. Meteor. Soc., Boston, MA, 392 pp.

Hess, G.D., Hicks, B.B. & Yamada, Y., 1981: The impact of the Wangara experiment. *Bound.-Layer Meteor.* **20**, 135–174.

Högström, U., 1996: Review of some basic characteristics of the atmospheric surface layer. *Bound.-Layer Meteor.* **78**, 215–246.

Holtslag, A.A.M. & Nieuwstadt, F.T.M., 1986: Scaling the atmospheric boundary layer. *Bound.-Layer Meteor.* **36**, 201–209.

Hunt, J.C.R., 1982: Diffusion in the stable boundary layer. *Atmospheric Turbulence and Air Pollution Modelling*, F. Nieuwstadt & H. van Dop, Eds., D. Reidel Publishing, Boston, MA, 231–274.

Hunt, J.C.R., Kaimal, J.C. & Gaynor, J.E., 1985: Some observations of turbulence in stable layers. *Quart. J. Roy. Meteor. Soc.* **111**, 793–815.

Izumi, J.S. & Caughey, S.J., 1976: Minnesota 1973 atmospheric boundary layer experiment data report. AFCRL-TR-76-0038, AFGL, Bedford, MA, 28 pp.

Kaimal, J.C., 1973: Turbulence spectra, length scales and structure parameters in the stable surface layer. *Bound.-Layer Meteor.* **4**, 289–309.

Kaimal, J.C. & Finnigan, J.J., 1994: *Atmospheric Boundary Layer Flows.* Oxford University Press, New York, 289 pp.

Kemp, J.R. & Thomson, D.J., 1996: Dispersion in stable boundary layers using large-eddy simulation. *Atmos. Environ.* **30**, 2911–2923.

Kim, J. & Mahrt, L., 1992: Simple formulation of turbulent mixing in the stable free atmosphere and nocturnal boundary layer. *Tellus* **44A**, 381–394.

Kondo, J., Kanechika, O. & Yasuda, N. 1978: Heat and momentum transfer under strong stability in the atmospheric surface layer. *J. Atmos. Sci.* **35**, 1012–1021.

Kraus, H., Malcher, J. & Schaller, E., 1985: A nocturnal low level jet during PUKK. *Bound.-Layer Meteor.* **31**, 187–195.

Kristensen, L., Jensen, N.O. & Peterson, E.L., 1981: Lateral dispersion of pollutants in a very stable atmosphere. *Atmos. Environ.* **15**, 837–844.

Lacser, A. & Arya, S.P.S., 1986: A comparative assessment of mixing-length parameterizations in the stably stratified nocturnal boundary layer (NBL). *Bound.-Layer Meteor.* **36**, 53–70.

Leahey, D.M., Hansen, M.C. & Schroeder, M.B., 1988: An analysis of wind fluctuation statistics collected under stable atmospheric conditions at three sites in Alberta, Canada. *J. Appl. Meteor.* **27**, 774–777.

Leahey, D.M., Hansen, M.C. & Schroeder, M.B., 1994: Variations of wind fluctuations observed at 10 m over flat terrain under stable conditions. *J. Appl. Meteor.* **33**, 712–720.

Lenschow, D.L., Li, X.S., Zhu, C.J. & Stankov, B.B., 1988: The stably stratified boundary layer over the Great Plains. I: Mean and Turbulence Structure. *Bound.-Layer Meteor.* **42**, 95–121.

Lettau, H. & Davidson, B., 1957: *Exploring the Atmosphere's First Mile*, 2 vols., Eds. Pergamon Press, New York.

Luhar, A.K. & Britter, R.E., 1989: A random walk model for dispersion in inhomogeneous turbulence in a convective boundary layer. *Atmos. Environ.* **23**, 1911–1924.

Luhar, A.K. & Rao, K.S., 1993: Random-walk model studies of the transport and diffusion of pollutants in katabatic flows. *Bound.-Layer Meteor.* **66**, 395–412.

Luhar, A.K. & Rao, K.S., 1994: Lagrangian stochastic dispersion model simulations of tracer data in nocturnal flows over complex terrain. *Atmos. Environ.* **28**, 3417–3431.

Mahrt, L., Heald, R.C., Lenschow, D.H., Stankov, B.B., & Troen, I.B., 1979: An observational study of the structure of the nocturnal boundary layer. *Bound.-Layer Meteor.* **17**, 247–264.

Mahrt, L., 1985: Vertical structure and turbulence in the very stable boundary layer. *J. Atmos. Sci.* **42**, 2333–2349.

Mahrt L. & Gamage, N., 1987: Observations of turbulence in stratified flows. *J. Atmos. Sci.* **44**, 1106–1121.

Mahrt, L., 1989: Intermittency of atmospheric turbulence. *J. Atmos. Sci.* **46**, 79–95.

Mason, P.J. & Derbyshire, S.H., 1990: Large-eddy simulation of the stably-stratified atmospheric boundary layer. *Bound.-Layer Meteor.* **53**, 117–162.

Mason, P.J. & Thomson, D.J., 1992: Stochastic backscatter in large-eddy simulation of boundary layers. *J. Fluid Mech.* **242**, 1561–1572.

Miles, J.W., 1961: On the stability of heterogeneous shear flow. *J. Fluid Mech.* **10**, 496–508.

Nappo, C.J. & Rao, K.S., 1987: A model study of pure katabatic flows. *Tellus* **39A**, 61–71.

Nappo, C.J., Rao, K.S. & Herwehe, J.A., 1989: Pollutant transport and diffusion in katabatic flows. *J. Appl. Meteor.* **28**, 617–625.

Nappo, C.J., 1991: Sporadic breakdowns of stability in the PBL over simple and complex terrain. *Bound.-Layer Meteor.* **54**, 69–87.

Nappo, C.J., 1992: Gravity-wave-generated turbulence and diffusion in the stable planetary boundary layer. *Air Pollution Modeling and its Applications IX*, H. van Dop & G. Kalos, eds., Plenum Press, New York.

Nappo, C.J. & Chimonas, G., 1992: Wave exchange between the ground surface and a boundary layer critical level. *J. Atmos. Sci.* **49**, 1075–1091.

Nieuwstadt, F.T.M. & Driedonks, A.G.M., 1979: The nocturnal boundary layer. A case study compared with model calculations. *J. Appl. Meteor.* **18**, 1397–1405.

Nieuwstadt, F.T.M. & Tennekes, H., 1981: A rate equation for the nocturnal boundary layer height. *J. Atmos. Sci.* **38**, 1418–1428.

Nieuwstadt, F.T.M. & van Dop, H., 1982: *Atmospheric Turbulence and Air Pollution Modelling*, D. Reidel, Boston, MA, 358 pp.

Nieuwstadt, F.T.M., 1984: The turbulent structure of the stable, nocturnal boundary layer. *J. Atmos. Sci.* **41**, 2202–2216.

Panofsky, H.A., Egolf, C.A. & Lipschutz, R., 1978: On characteristics of wind direction fluctuations in the surface layer. *Bound.-Layer Meteor.* **15**, 439–446.

Panofsky, H.A. & Dutton, J.A., 1984: *Atmospheric Turbulence.* John Wiley & Sons, New York, NY, 397 pp.

Parish, T.R., 1988: The surface windfield over the Antarctic continent. *Rev. Geophys.* **26**, 169–180.

Pasquill, F. & Smith, F.B., 1983: *Atmospheric Diffusion*, 3rd Ed. Wiley & Sons, New York, 437 pp.

Pearson, H.J., Puttock, J.S. & Hunt, J.C.R., 1983: A statistical model of fluid-element motions and vertical diffusion in a homogeneous stratified turbulent flow. *J. Fluid Mech.* **120**, 219–249.

Phillips, P. & Panofsky, H.A., 1982: A re-examination of lateral dispersion from continuous sources. *Atmos. Environ.* **16**, 1851–1860.

Rao, K.S. & Snodgrass, H.F., 1978: The structure of the nocturnal boundary layer. ATDL contribution 78/9, NOAA, Oak Ridge, TN, 44 pp.

Rao, K.S. & Snodgrass, H.F., 1979: Some parameterizations of the nocturnal boundary layer. *Bound.-Layer Meteor.* **17**, 15–28.

Rao, K.S. & Snodgrass, H.F., 1981: A nonstationary nocturnal drainage flow model. *Bound.-Layer Meteor.* **20**, 309–320.

Rao, K.S., Singh, M.P. & Ghosh, S., 1986: The Bhopal gas tragedy. *Determination of Atmospheric Dilution for Emergency Preparedness*, Joint EPA-DOE Technical Workshop, Research Triangle Park, NC. ATDD contribution 86/24, Oak Ridge, TN.

Rao, K.S. & Schaub, M.A., 1990: Observed variations of σ_θ and σ_ϕ in the nocturnal drainage flow in a deep valley. *Bound.-Layer Meteor.* **51**, 31–48.

Raynor, G.S. & Hayes, J.V., 1984: Wind direction meander at a coastal site during onshore flow. *J. Clim. Appl. Meteor.* **23**, 967–978.

Rodi, W., 1985: Calculation of stably stratified shear-layer flows with a buoyancy-extended $\bar{e} - \epsilon$ turbulence model. *Turbulence and Diffusion in Stable Environments*, J.C.R. Hunt, Ed., Clarendon Press, Oxford, 111–140.

Sawford, B.L., 1993: Recent developments in the Lagrangian stochastic theory of turbulent dispersion. *Bound.-Layer Meteor.* **62**, 197–215.

Sethuraman, S., 1980: A case of persistent breaking of internal gravity waves in the atmospheric surface layer over the ocean. *Bound.-Layer Meteor.* **19**, 67–80.

Sharan, M., McNider, R.T., Gopalakrishnan, S.J. & Singh, M.P., 1995: Bhopal gas leak: A numerical simulation of episodic dispersion. *Atmos. Environ.* **29**, 2061–2074.

Sharan, M., Singh, M.P. & Yadav, A.K., 1996: Mathematical model for atmospheric dispersion in low winds with eddy diffusivities as linear functions of downwind distance. *Atmos. Environ.* **30**, 1137–1145.

Singh, M.P. & Ghosh, S., 1987: Bhopal gas tragedy: model simulation of dispersion scenario. *J. Hazard. Mater.*, **17**, 1-22.

Sorbjan, Z., 1986: On similarity in the atmospheric boundary layer. *Bound.-Layer Meteor.* **34**, 377–397.

Sorbjan, Z., 1987: An examination of local similarity theory in the stably stratified boundary layer. *Bound.-Layer Meteor.* **38**, 63–71.

Sorbjan, Z., 1989: *Structure of the Atmospheric Boundary Layer*. Prentice Hall, Englewood Cliffs, NJ, 317 pp.

Smedman, A., 1988: Observations of a multi-level turbulence structure in a very stable atmospheric boundary layer. *Bound.-Layer Meteor.* **44**, 231–253.

Smedman, A., 1991: Some turbulence characteristics in stable atmospheric boundary layer flow. *J. Atmos. Sci.* **48**, 856–868.

Stull, R.B., 1988: *An Introduction to Boundary Layer Meteorology*. Kluwer Academic Publishers, Boston, 666 pp.

Surridge, A.D., 1986: The evolution of the nocturnal temperature inversion. *Bound.-Layer Meteor.* **36**, 295–305.

Taylor, G.I., 1921: Diffusion by continuous movements. *Proc. London Math. Soc.*, **20**, 196–202.

Thomson, D.J., 1987: Criteria for the selection of stochastic models of particle trajectories in turbulent flow. *J. Fluid Mech.* **180**, 529–556.

Tjemkes, S.A. & Duynkerke, P.G., 1989: The nocturnal boundary layer: model calculations compared with observations. *J. Appl. Meteor.* **28**, 161–175.

Van Ulden, A.P. & Holtslag, A.A.M., 1985: Estimation of atmospheric boundary layer parameters for diffusion calculations. *J. Clim. Appl. Meteor.* **24**, 1196–1207.

Venkatram, A., 1988: Dispersion in the stable boundary layer. *Lectures on Air Pollution Modeling*, A. Venkatram & J. C. Wyngaard, Eds., Amer. Meteor. Soc., Boston, MA, 229–265.

Venkatram, A. & Wyngaard, J.C., 1988: *Lectures on Air Pollution Modeling*, Amer. Meteor. Soc., Boston, MA, 390 pp.

Weil, J.C., 1988: Atmospheric dispersion–observations and models. *Flow and Transport in the Natural Environment: Advances and Applications.* W.L. Steffen & O.T. Denmead, Eds., Springer-Verlag, 352–376.

Wilson, J.D. & Sawford, B.L., 1996: Review of Lagrangian stochastic models for trajectories in the turbulent atmosphere. *Bound.-Layer Meteor.* **78**, 191–210.

Wyngaard, J.C., 1973: On surface layer turbulence, Chapter 3 in *Workshop on Micrometeorology*, D.A. Haugen, Ed., Amer. Meteor. Soc., Boston, MA, 101–149.

Wyngaard, J.C., 1975: Modeling the planetary boundary layer–extension to the stable case. *Bound.-Layer Meteor.* **9**, 441–46.

Yamada, T. & Mellor, G., 1975: A simulation of the Wangara atmospheric boundary-layer data. *J. Atmos. Sci.* **32**, 2309–2329.

Yu, T.W., 1978: Determining the height of the nocturnal boundary layer. *J. Appl. Meteor.* **17**, 28–33.

Zannetti, P., 1986: A new mixed segmented-puff approach for dispersion Modelling. *Atmos. Environ.* **20**, 1121–1130.

Zeman, O., 1979: Parameterization of the dynamics of stable boundary layers and nocturnal jets. *J. Atmos. Sci.* **36**, 792–804.

Zilitinkevich, S.S., 1972: On the determination of the height of the Ekman boundary layer. *Bound.-Layer Meteor.* **3**, 141–145.

Zilitinkevich, S.S., 1975: Resistance laws and prediction equations for the depth of the planetary boundary layer. *J. Atmos. Sci.* **32**, 741–752.

Zilitinkevich, S.S., 1989: Velocity profiles, the resistance law and the dissipation rate of mean flow turbulent kinetic energy in a neutrally and stably stratified planetary boundary layer. *Bound.-Layer Meteor.* **46**, 367–387.

Chapter 2
Turbulence and dispersion in the stable atmospheric boundary layer
Section B

R.T. McNider[a], M.P. Singh[b], & S.Gupta[c]

[a]*Department of Mathematics, University of Alabama in Huntsville, Huntsville, AL 35899, USA*
[b]*Earth System Science Laboratory, University of Alabama in Huntsville, Huntsville, AL 35899, USA*
[c]*Atmospheric Science Department, University of Alabama in Huntsville, Huntsville, AL 35899, USA*

Abstract

Two important aspects relevant in the context of stable boundary layer (SBL) are highlighted here.

1. Techniques of nonlinear dynamics and stability analysis have been used to show the existence of multivalued solutions for certain values of external parameters which have strong implications for the predictability of the SBL.

2. Vertical shear in the horizontal wind produced by diurnal and/or inertial oscillations (IO) could maintain plume growth rates consistent with the observations, which are nearly linear with diffusion times. Radar profiler technology has been used to carry out a time series analysis of the energy spectra based on which results from a recently conducted air pollution field study are found to give better results when compared with observations.

1   Introduction

The nocturnal atmospheric boundary layer evolves through a competition between the suppression of turbulent energy by radiative cooling and generation of turbulent energy by shear. Unlike in the unstable boundary layer where structure and turbulent intensity change only by degree as a function of external parameters such as geostrophic wind speed or surface roughness, in the stable boundary layer variations in parameters can tilt the outcome of the competition to produce distinctly different mean and turbulent structures. These variations can then have a major impact on parameters such as surface temperature important to agricultural interests or the dispersion and transport of pollutants. In the following we first emphasize the development of the stable boundary layer in terms of these competing processes and the implications for predictability. Second, we examine the implications of the differing behavior to the support of inertial oscillations and their impact on medium range transport and diffusion.

2   Predictability of the Stable Boundary Layer

Concerns about the limits of predictability of the atmosphere have long spurred theoretical and numerical studies of the fundamental primitive equations and simpler analogues (see e.g. Lorenz (1963), Lorenz (1982)). These studies exposed the extremely complicated behavior that might be embedded within the full equations. They showed that even modest nonlinearities in the type of equations employed in atmospheric predictions displayed a sensitive dependence on initial conditions and could yield totally different solutions. In numerical models such variations could arise through truncation round-off error so that in terms of traditional numerical analysis the problem is characterized as ill-posed. Recently, McNider *et al.* (1995) showed that a simplified set of equations governing the stable boundary layer also contained nonlinearities that can lead to mutiple solutions and thus unpredictabiity. The following summarizes some of the results of this study and newer analyses using more complete mathematical representation of the physical system.

As mentioned above the nocturnal stable boundary layer is dynamically interesting because of the competing processes of growing thermal stability due to surface and radiational cooling and the development of vertical shear in the horizontal winds. Using concepts of Richardson Number instability (see Businger, (1973)), if the thermal stability develops faster than the shear then the surface can become decoupled from the layers above leading to light winds at the surface as vertical turbulent transfer of momentum is reduced. Rapid cooling of the surface also occurs as heat transfer from above is suppressed. This situation can produce low surface temperatures and is relevant in frost prediction.

On the other hand, if shear develops faster than the thermal stability, then high levels of turbulence can be supported. This maintains momentum and heat flux to the surface leading to warmer and windier conditions near the surface. Such conditions are important for mixing of elevated pollutant plumes to the surface.

To demonstrate these differing states, Fig. 1 shows the results from a mesoscale model run in a one-dimensional mode for the nocturnal boundary layer. The model is described by McNider and Pielke (1981) and incorporates local Richardson Number closure developed by Blackadar (1979). This mesoscale model contains closure, physics and resolution typical of today's generation of mesoscale and regional models. The figure shows temperature prediction at a height of four meters as a function of time for three different imposed geostrophic wind speeds.

The temperature trace for the low geostrophic wind speeds shows the characteristics of the decoupled case in which the temperature drops exponentially over the time period of integration. The trace for the high wind speed case shows an initial, almost parabolic, cooling, but, then levels off to a constant and much warmer solution than the low wind speed case.

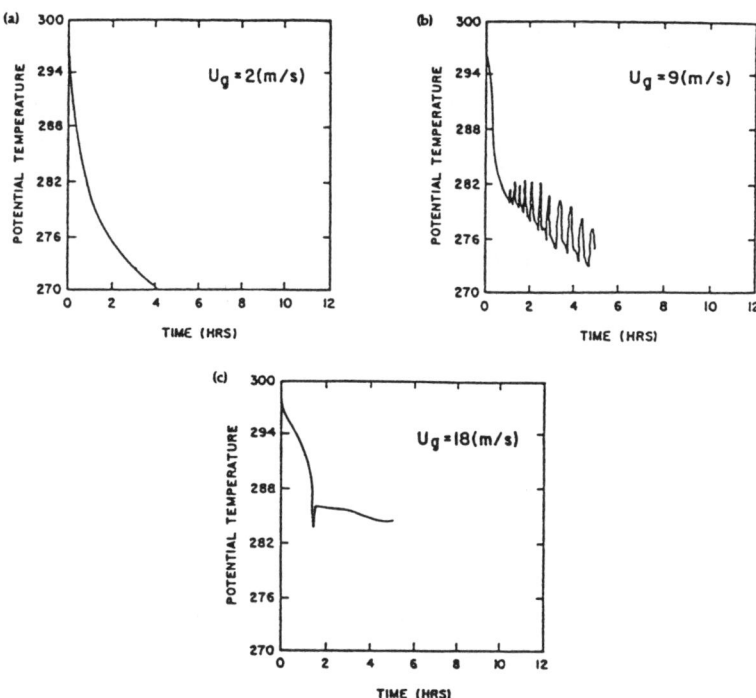

Figure 1: Time-dependent solutions for temperature from a mesoscale boundary layer model for different imposed geostrophic velocities. (Reprinted from *J. Atmos.Sci.*, 1995, McNider, *et al.)*

Some investigators (e.g. Businger, 1973) have postulated that oscillatory behavior may be produced when shear, in the face of thermal stability, eventually builds to the point that a breakdown in the boundary layer occurs. Then as the shear is removed by vertical mixing a renewed cycle of cooling can reintroduce the thermal stability. Revelle (1993) using a simple representation of the cooling boundary layer produced cyclical solutions and Blackadar (1979) for a particular set of parameters found oscillatory behavior in a one dimensional boundary layer model. Interestingly, in Fig. 1, an intermediate geostrophic wind case shows an oscillatory behavior superimposed on a linear cooling trend.

2.1   Analysis of the Stable Boundary Layer Equations

Whether the surface layer decouples or remains turbulently coupled to the air above is an important practical question in weather forecasting or air pollution modeling. What combination of external parameters leads to one or the other state? Is there an oscillatory regime? How predictable and stable is the selection of states?

2.2   Analysis of a Simple System of Two Layer Ordinary Differential Equations

McNider *et al.* (1995) began with a set of boundary layer equations of the type used in current regional scale forecast models and then simplified the equations to a set of ordinary differential equations by truncating the system to two layers (Fig. 2).

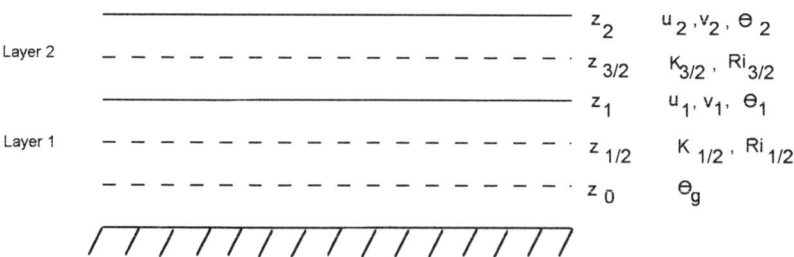

Figure 2: Schematic of coordinate system used in the present study. (Reprinted from *J. Atmos.Sci.*, 1995, McNider, *et al.*)

As a simplification we initially neglect clear-air radiative cooling, *i. e.*, $R_c(\theta) = 0$. We now turn our attention to a finite-difference representation of these equations and use the two layer model (Fig. 2) to get the wind speed

$u_1 = u(z_1), v_1 = v(z_1)$ and the potential temperature $\theta_1 = \theta(z_1)$, at the first level $(z = z_1), u_2 = u(z_2), v_2 = v(z_2), \theta_2 = \theta(z_2)$, at the second level $(z = z_2)$ and θ_g at the ground level $(z=0)$. With the truncation to the two layer system we reach the following system of four first-order ordinary differential equations (ODE):

$$\frac{du_1}{dt} = f(v_1 - v_G) + \frac{1}{z_{3/2} - z_{1/2}}\left(\frac{K_{m3/2}(u_2 - u_1)}{z_2 - z_1} - u_*^2 \cos(\psi)\right) \tag{1a}$$

$$\frac{dv_1}{dt} = f(u_G - u_1) + \frac{1}{z_{3/2} - z_{1/2}}\left(\frac{K_{m3/2}(v_2 - v_1)}{z_2 - z_1} - u_*^2 \sin(\psi)\right) \tag{1b}$$

$$\frac{d\theta_1}{dt} = \frac{1}{z_{3/2} - z_{1/2}}\left(\frac{K_{h3/2}(\theta_2 - \theta_1)}{z_2 - z_1} - u_* \theta_*\right) \tag{1c}$$

$$\frac{d\theta_g}{dt} = \frac{1}{C_g}\left(I_\downarrow - \sigma\theta_g^4 - H_0\right) - \kappa_m\left(\theta_g - \theta_m\right) \tag{1d}$$

where in this set $u=u(t,z)$, $v=v(t,z)$, $\theta = \theta(t,z)$ and $\theta_g = \theta_g(t)$ are the east-west, north-south velocity components, potential temperature, and ground temperature, respectively. The initial conditions are taken to be

$$u(0,z) = u_0, \ v(0,z) = v_0, \ \theta(0,z) = \theta_0, \ \theta_g(0) = \theta_{g0},$$

and the relevant boundary conditions, for $t \ge 0$, are

$$u(t,z_2) = u_G, \ v(t,z_2) = v_G, \ \theta(t,z_2) = \theta_2,$$

where

$$z_{1/2} = z_1/2, \ z_{3/2} = (z_1 + z_2)/2, \ K_{3/2} = f(Ri_{3/2})l_{3/2}^2 s_{3/2}.$$

Following Blackadar (1979), the exchange coefficients for momentum and heat are taken to have the forms

$$K_m = f_m(Ri)l^2 s, \qquad K_h = f_h(Ri)l^2 s$$

where the Richardson number formulations for the momentum and heat stability functions are given by

$$f_m(Ri) = f_h(Ri) = f(Ri) = \begin{cases} (1 - Ri/Ri_c)^2 & Ri < Ri_c \\ 0 & Ri \geq Ri_c \end{cases}$$

$$(2)$$

and $Ri_c = 0.25$ is the critical Richardson number. The quadratic form for the stability function was developed by England and McNider (1995) and differs from the linear form developed by Blackadar. The quadratic form is consistent with boundary-layer theory and also has the advantage of continuity at the critical value. The Richardson number Ri has the form:

$$Ri = \frac{g}{\theta_A} \frac{\partial \theta}{\partial z} \Big/ \left(\left(\frac{\partial u}{\partial z} \right)^2 + \left(\frac{\partial v}{\partial z} \right)^2 \right),$$

The Coriolis parameter is given by f, while u_G and v_G are the zonal and meridional components of the geostrophic wind in a horizontal plane. The exchange coefficients for momentum and heat are given by K_m and K_h, respectively.

The symbol C_g is the heat capacity per unit area ($C_g = 14600$ $JK^{-1}m^{-1}$ for dry peat soil). The longwave back radiation from the atmosphere, $I_\downarrow$, based upon the form used by Staley and Jurica (1972) is given by:

$$I_\downarrow = \sigma \left(Q_c + 0.67(1 - Q_c)(1670 Q_a)^{0.08} \right) \theta_1^4,$$

$$(3)$$

where σ is the Stefan-Boltzmann constant, Q_c is the cloud fraction ($Q_c = 0$ for a clear sky), Q_a is the specific humidity at the height z_a, H_0 is the heat flux carried away from the surface by turbulence, ρ is the air density and C_p is the specific heat at constant pressure. The soil transfer coefficient is given by κ_m, and θ_m is the mean temperature of the substrate and is taken as the mean temperature of the surface air during the most recent 24 hours (see Blackadar (1979) and McNider et al. (1995) for further explanation).

The set of equations (1a-d) can be considered a nonlinear dynamical system with the nonlinearities entering through the turbulent diffusion terms. Such nonlinearities effectively prevent analytical solutions; however, the behavior of the solutions can be quantitatively examined as a function of external parameters using methods of numerical bifurcation theory. Thus, despite the lack of a closed-form solution, parameter dependence and characteristics of the solution can be developed.

Techniques of nonlinear analysis have expanded in recent years using numerical continuation methods. Sophisticated software packages are also now readily available to carry out the analysis. For those not familiar with numerical continuation, Seydel (1988) and Doedel and Kernevéz (1986) provide an excellent overview to practical techniques of nonlinear analysis and continuation (see McNider *et al.* (1995) for an introduction to this technique).

We next use numerical continuation to analyze the system (1a-d) and characterize the behavior of the system as a function of external parameters. In the analysis we choose the external parameters to be geostrophic wind, U_g, roughness length, z_0, layer heights, z_1 and z_2, and the Coriolis parameter, f. In order to carry out the analysis of the system of equations (1), a software analysis package, *AUTO* (Doedel and Kernevéz (1986)), was utilized.

In order to carry out the bifurcation analysis, starting steady state solutions for $\left(u_s, v_s, \theta_s \text{ and } \theta_{gs}\right)$ are needed for the continuation. While this can normally be accomplished directly in *AUTO*, the length of computational time required to carry out the integration of over ten hours of model simulation using the high order of numerical accuracy in *AUTO* was cumbersome. To obtain initial equilibrium solutions we utilized a fourth-order Runge-Kutta technique for the coupled set to obtain steady-state solutions as a limit for large times, of time-dependent solutions. The resulting equilibria were used as the starting points in *AUTO* for the bifurcation analysis.

The first results for temperature are seen in Fig. 3 in which the geostrophic wind speed, u_G, is chosen as the bifurcation parameter. The four panels show the bifurcation diagram for four different roughness lengths, z_0. We will later use z_0 as a bifurcation parameter. In the first panel $z_0 = 0.1$ m, we see that a rapid change occurs in the solution near $u_G = 13\ ms^{-1}$. This is the transition from the tranquil state to the left, in which the surface is decoupled from the layer above, to the turbulently connected state to the right. This is consistent with physical intuition and experience in the stable boundary layer. For high geostrophic speeds we would expect shear-generated turbulence to cause the surface to remain relatively warm as turbulent heat flux counteracts the radiative cooling. At lower geostrophic wind speeds the surface layer cools to the point that thermal stability impedes turbulent transfer, leading to a rapid cooling of the surface.

For roughness lengths smaller than 0.1 m, the transition is gradual. While the above appears to be physically realistic and is instructive, the solutions versus the bifurcation parameter remain single-valued and are uninteresting from the point of view of dynamical analysis in that there are no special turning points or bifurcations. As roughness lengths are increased, the bifurcation diagrams change. In the second panel for $z_0 = 0.5$m, limit points or turning points are found and the solution becomes multivalued for the bifurcation parameter. Points labeled as 2 and 3 are stable limit points. There is also an unstable solution region indicated by the dashed line. Stable attracting solutions are indicated by the solid lines.

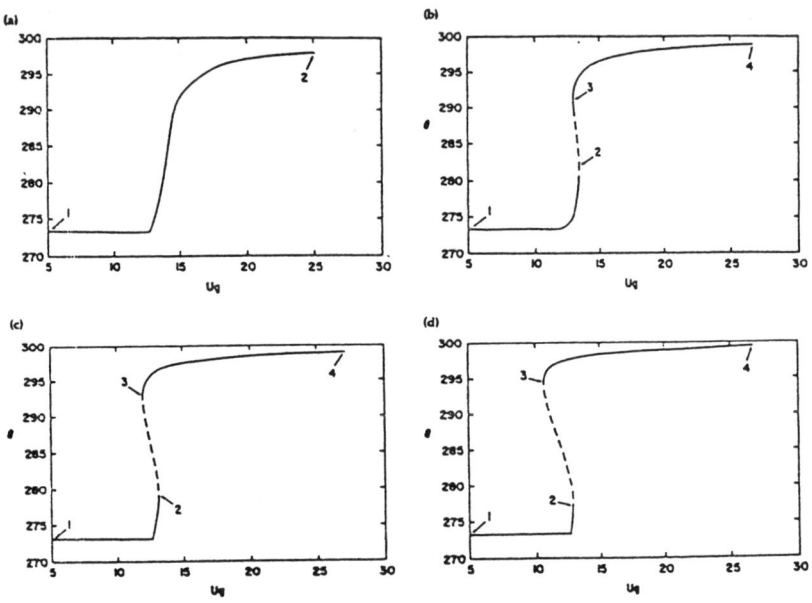

Figure 3. Bifurcation diagram for temperature versus the bifurcation
parameter geostrophic wind speed u_G for four different values of
z_0: (a) $z_0 = 0.1m$, (b) $z_0 = 0.5m$, (c) $z_0 = 1.0m$, and (d) $2.0m$. Points 1 and
4 are starting and ending points, while 2 and 3 are stable limit points.
Stable solution regions are given by solid lines, and unstable solution
regions by dashed lines.
(Reprinted from *J. Atmos. Sci.*, 1995, McNider, *et al.*)

In the third and fourth panels, as the roughness length is increased to 1 m and 2 m, respectively, the solution curve becomes even steeper, leading to multivalued solutions for a larger range of geostrophic winds. The length of the unstable region has also increased. Note the substantial difference in temperature between the surface-decoupled regime to the left and the turbulently connected regime to the right.

The dependence of the solution on roughness length is perhaps expected in that greater roughness lengths increase momentum transfer to the surface by increasing u_*. This in turn increases the shear between the two layers maintaining Richardson numbers at lower values. In terms of the closure, this in turn maintains larger diffusivity values connecting the two layers.

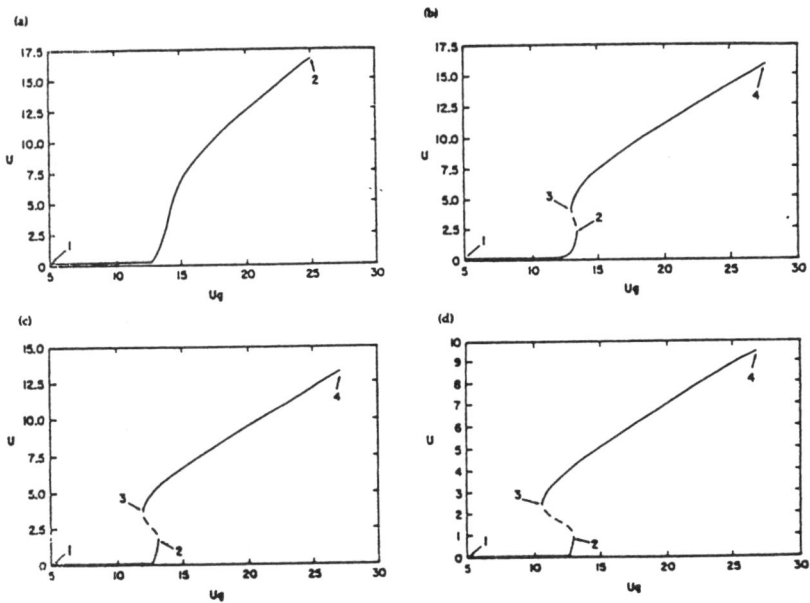

Figure 4. Same as Fig. 3 except for wind speed.
(Reprinted from *J. Atmos.Sci.*, 1995, McNider, *et al.*)

Figure 4 shows a similar panel of results for wind speed, u, in the lower layer. As roughness increases, bifurcation points are introduced, leading to multivalued solutions in a critical window of geostrophic wind speeds. Once the surface layer becomes turbulently connected to the outer layer to the right of limit point 3, the wind speed at the turbulently connected to the outer layer to the right of limit point 3, the wind speed at the turbulently connected to the outer layer to the right of limit point 3, the wind speed at the surface behaves approximately linearly with respect to geostrophic wind speed. The results to this analysis have practical importance to the predictability of the stable boundary layer. Because of the multivalued solution regime, even slight changes in initial conditions may change the limiting solution as time increases. While the results given here are for the discrete system and show dependence on the layer depth, the discrete analog is very close to the PDE system which is solved in weather-forecast and mesoscale models. The results show that the solution of stable boundary layer equations may be indeterminate for certain ranges of imposed geostrophic winds. Practically, this means that frost prediction or pollutant dispersion cannot be made with confidence in certain parameter regions. If this type of behavior holds in the full PDE system, it also implies that the addition of physics or numerical sophistication in the system may be irrelevant in improving predictability.

It should be noted the results presented here analyze the behavior of a model of the atmosphere. The closure assumptions and other simplifications in the model may make the analyzed behavior of this system nontransferable to the real atmosphere. However, it does provide a framework to attempt to understande the limits of predictability of the real atmosphere and a practical analysis of the behavior of actual numerical models which are used to make weather predictions.

2.2.1   Analysis of System of Multi-Layer Partial Differential Equations

The analysis of the two layer system above showed that a region of multiple solutions exists which has practical implications to forecasting. The key question is whether this behavior holds in the PDE set including full z dependence. To address this question the PDE system is discretized into a multi-layer model using the orthogonal collocation method and the techniques of numerical continuation and nonlinear stability analysis (Shi, 1997). Also, a more realistic boundary condition has been incorporated by assuming a balance between the radiative cooling and turbulent heat flux in the steady state as the lower boundary condition at z=zo (in place of $\theta=\theta g$). It is found that the system of ordinary differential equations incorporating higher resolution has similar S-shaped bifurcation diagrams (Fig.5), and range of values of bifurcation parameters in which multiple solutions exist, thereby confirming the results obtained from the simpler two-layer model regarding the limitations of predictability of stable boundary layer discussed in the preceding section. The multiple solution regime is sensitive to the depth of the atmosphere (i.e.,the top value z=Z at which the upper boundary condition is prescribed). It is found that as Z increases, the range of U_G (Geostrophic wind) in which multiple solutions exist, shifts to the right. Since intensity of turbulence is weak during the night, the depth to which the surface can connect as the night progresses can be of the order of the depth of the inversion layer which is taken to be about 200m.

The dependence of the solutions on the atmospheric depth is perhaps not unexpected. Because we are examining the steady solutions, the connectivity between the lower boundary conditions and upper boundary conditions are provided through the parobolic form of the differential equations. Thus, eventually a balance must develop and this balance is sensitive to the physical distance between the top and bottom boundary. The key factor is that the multiple solution regimes found in the two layer system above are also found in the PDE system.

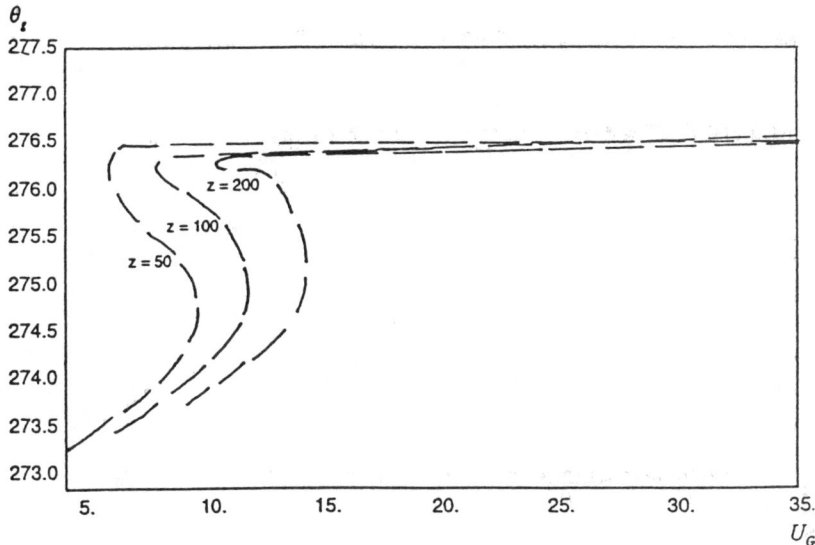

Figure 5. Bif. Diag. (lower B. L. at $z = z_0$, $a\left(\theta_g - \theta\right) - \dfrac{b\,H_0}{\rho c_p z_0} = 0$).

2.2.2 Oscillations and Intermittency in the nocturnal boundary layer (NBL)

While average levels of turbulence and diffusion in the stable boundary layer are generally low, sporadic outbreaks of turbulence and enhanced vertical diffusion are common in the NBL. These facts strongly affect the modeling of the atmospheric boundary layer. The mechanisms associated with boundary layer breakdowns are not fully understood.

Revelle (1993) conducted a theoretical study of the intermittent breakdowns of the NBL using a one-dimensional, flat terrain, boundary layer model based on Blackadar's studies (1979). It was found that the turbulent burst effects predominate in the lowest tens of meters of the NBL where an alternation between successive regimes of laminar and turbulent flow drive either rapid cooling or rapid heating, respectively, of the near-surface air temperature under certain conditions. The rather discontinuous change of the air temperature occurs with variable amplitudes at irregular intervals depending critically on the specified surface roughness length, the prevailing geostrophic wind, magnitude of the radiation cooling with respect to the soil and its thermal properties.

To understand the phenomenon of intermittency in the flow development, we re-examine Fig.1 in which intermediate geostrophic wind case shows an oscillatory behavior of the flow. One may be tempted to correlate the flow development in the three cases (Fig.1) with the nature of eigen values of the Jacobian matrix of the system (eqns 1a-d) at the (i) initial state of the flow, (ii) equilibrium state of the system. However, recent work (McNider et al. (1996)) shows that computation of eigen values does not provide any clue to the phenomenon of intermittency, which is, essentially, local in character. It is difficult, at this stage of research in this field, to lay down a mathematical criterion for predicting the existence of transient oscillations of moderate amplitude during the flow development. However, it has been found that, generally, the solutions in the neighborhood of the unstable side of the Hopf Bifurcation point are known to have periodic solutions with moderate amplitude (Seydel, (1988)). This does not imply that existence of Hopf Bifurcation point is a necessary condition for the intermittency in the flow development.

The aperiodicity of turbulence bursts found frequently in the NBL, as pointed out by Revelle (1993), is extremely sensitive to the initial conditions of the forcing. Besides considering the Surface Energy Budget equation, Revelle introduced the Thermodynamics Surface air equation in both of which effect of radiative cooling was considered. For certain range of parametric values (with $U_G\sim2$ m/s), he was able to get intermittency in the flow with period ranging from 3 - 4 hours.

3   Inertial Oscillations in the Stable Boundary Layer and Impact on Horizontal Dispersion

One characteristic of the solution above in which the surface becomes turbulently disconnected from the layer above is that the effective surface drag in the layer aloft is greatly reduced relative to its daytime value. As first described by Blackadar (1957), this layer when freed from the frictional restraints accelerates. In the rotating Earth coordinate system, this leads to an inertial oscillation as the fluid seeks a new frictional-geostrophic balance. The result is that this layer produces a jet like structure as it goes through the inertial oscillation. Considerable speed and directional shear is also generated between this layer and the surface layer which is still turbulently connected to the surface.

It has been recognized for many years that this inertial oscillation could affect air pollution transport and dispersion (Husar (1978)). Moran et al. (1987), McNider et al. (1993) and Singh et al. (1993) began to quantify the role of the inertial oscillation in affecting horizontal dispersion rates.

Observations (Carras and Williams (1981) and Gifford (1985)) indicate that horizontal plume growth rate in the atmospheric boundary layer for long travel distances is larger than the growth rate implied by an atmospheric energy spectrum with a mesoscale gap. This discrepancy led Gifford (1985) to speculate that mesoscale eddies might be responsible for the sustained growth. Using the flow fields derived from a dynamic numerical model together with a Lagrangian particle model to simulate particle trajectories, McNider et al. (1988) conducted experiments to evaluate the role that shear plays in dispersion in the planetary boundary layer (PBL) over time scales of 24-48 h. The experiments indicated that vertical shear in the horizontal wind produced by diurnal and/or inertial oscillations could maintain plume growth rates consistent with the observations, which are nearly linear with diffusion times.

The physical explanation put forward in the McNider et al. (1988) study is that horizontal dispersion in the PBL is controlled by the shear which develops overnight. This process proceeds as follows: during the daytime solar heating creates a deep convective boundary layer. Following sunset and during the night as the surface cools, a shallow nocturnal boundary layer develops below the old daytime boundary layer (hereafter the region between the nocturnal boundary layer and the old daytime boundary layer will be called the "residual layer" (see Fig.15)). The daytime boundary layer is well mixed, and the wind profiles are fairly uniform. The residual layer exhibits a marked acceleration to supergeostrophic speeds several hours after sunset and produces a significant low-level jet. This phenomena was first explained by Blackadar (1957), who argued that the low-level jet resulted from an undamped inertial oscillation of the ageostrophic component of the subgeostrophic wind where it was released from frictional constraints near sunset. This wind maximum occurs within the lowest kilometer of the atmosphere and often lasts for most of the night (Thorpe and Guymer (1977)). As the wind in the residual layer goes through the inertial oscillation, McNider et al. (1988) showed that the resulting shear causes the effective plume width to increase substantially. The role of the inertial oscillation in affecting horizontal diffusion is a two step process: the vertical shear in the horizontal induced by the inertial oscillation first distorts the shape of the plume and then subsequent vertical mixing the next morning operates on this distorted plume to produce a wider plume than would have occurred in the absence of the I.O. Figures 6 (a and b) show a schematic of this process. True diffusion does not occur since local changes in concentration in the nighttime tilted plume are small. As the sun comes up, the plume is mixed vertically. At this point the overnight shear affects local concentration in that the plume volume is large and concentrations are reduced dramatically. The distortion of the plume by the mean wind shear in the inertial oscillation and then subsequent mixing is an example of a delayed diffusion evidently first discussed by Pasquill (1962).

In McNider et al. (1988), a numerical boundary layer model coupled with a Lagrangian Particle Model was used to examine quantitatively the role of the inertial oscillation in affecting long range dispersion rates. Figure 7 shows a sample of how microscale turbulence, inertial oscillation induced boundary layer shear and geostrophic shear contribute to plume spread.

A series of remarkable long range plume dispersion experiments had been made from the Mt. Isa in the relatively clean background air over Australia (see Carras and Williams (1981)). Figures 8 (a-c) show the results of the model experiments plotted against the Australian data. The depiction shows that inclusion of the inertial oscillation can effectively explain the continued spread of the plume at a rate proportional to time out mesoscale transport distances. The model simulations also showed that a period of accelerating diffusion ocurred overnight consistent with a $t^{3/2}$ growth rate in the observed data noted by Gifford (1983).

Since the role of the inertial oscillation makes such a large impact on plume dispersion rates and the fact that the inertial oscillation is dependent upon latitude it seems a natural extension to examine the effect of latitude on plume dispersion. Such a study was undertaken by McNider et al. (1993) in which series of numerical experiments was undertaken using the 1D version of the mesoscale model and the Lagrangian particle model to examine the role of external parameters such as latitude and surface roughness in controlling lateral dispersion. A generic sounding was used to initiate the thermodynamic fields in the model. The mesoscale model was integrated in time for 36 h starting at sunrise. Table 1 gives the parameters for the baseline experiment.

Table 1. Mesoscale model parameters for horizontal dispersion experiments.
(Reprinted from *J. Atmos. Sci* . McNider, *et al.*, 1995*)*

Number of vertical levels	26
	(5, 30, 80, 150, 208, 445, 690, 943, 1203, 1470, 1743, 2023, 2311, 2609, 2916, 3232, 3558, 3895, 4245, 4607, 4983, 5375, 5784, 6212, 6661, 7136 m)
Latitude	45N
Surface roughness	10cm
Geostrophic wind speed	4 ms-1
Geostrophic wind direction	90 (easterly)
Albedo	0.2
Soil Conductivity	0.0018 W cm^{-1} K^{-1}
Soil density	1.4 g cm^{-3}
Soil specific heat	0.37 cal g^{-1} K^{-1}
Soil wetness	0.05
Time step	3 s
Day of the year	1-Apr
Initial Potential Temp.	298 K

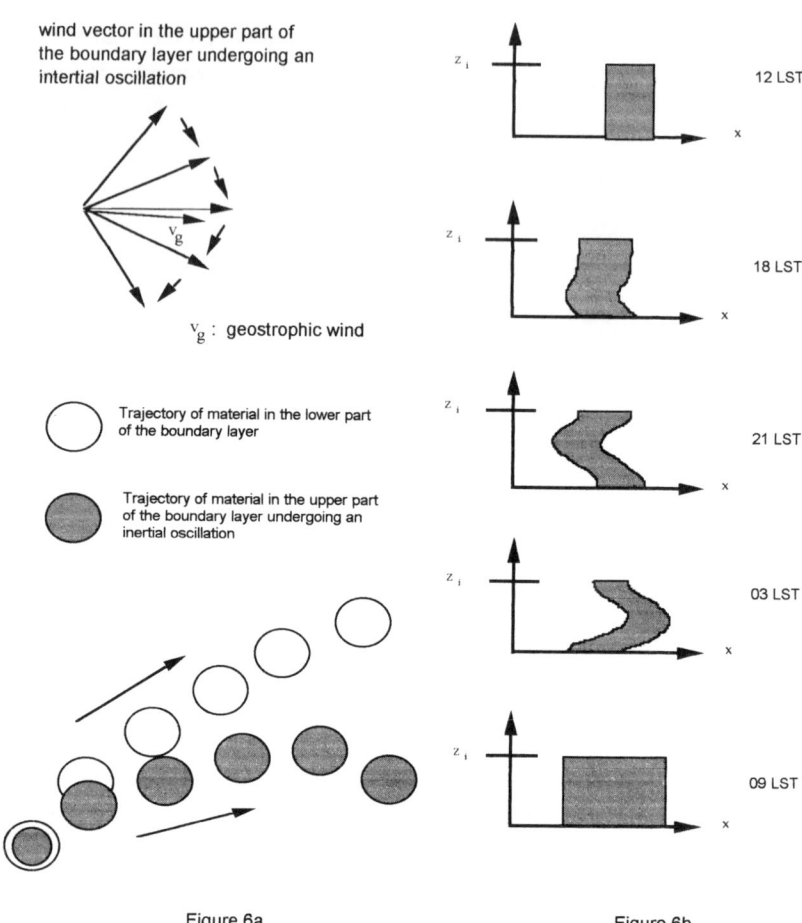

Figure 6a

Figure 6b

Figure 6a. Schematic showing an inertial oscillation and corresponding effect on horizontal dispersion.

Figure 6b. Schematic showing the effect of shear on the vertical distribution and horizontal displacement of pollutants. The schematic gives a view looking downwind. Initially the pollutant is well mixed up to the boundary layer depth. After 0600 LST the inertial oscillation begins to distort the plume. If the next morning's mixing catches the intertial oscillation in the current phase, the result is a very wide plume as seen at 0900 LST. (Reprinted from *Atmos. Environ.*, 1993, McNider, *et al.* with kind permission from Elsevier Science Ltd.)

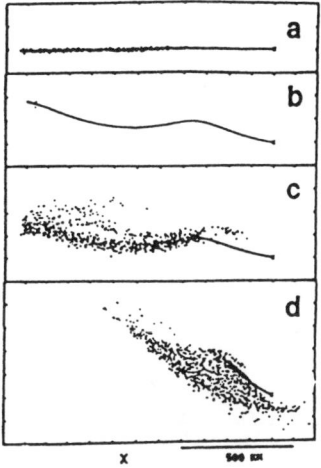

Figure 7. Instantaneous depiction of plume release after 42 h. (a) Only PBL turbulence but no PBL shear and no Coriolis force. (b) Including Coriolis force but no PBL turbulence. (c) Includes PBL turburlence, PBL shear and Coriolis force. (d) Same as (c) except for baroclinic case which includes geostrophic shear.
(Reprinted from *Atmos. Environ.*, 1988, McNider, *et al.* with kind permission from Elsevier Science Ltd.)

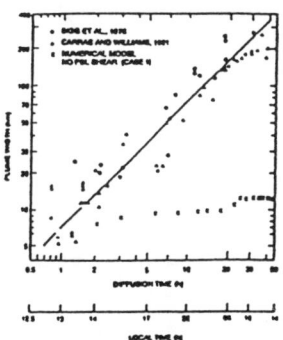

Figure 8a. Horizontal spread for a short-term release (pseudo-puff) into barotropic flow. Vertical shear of the horizontal wind was not included so dispersion is totally due to PBL-scale turbulence. Local time applies to meteorological simulation results.
(Reprinted from *Atmos. Environ.*, 1988, McNider, *et al.* with kind permission from Elsevier Science Ltd.)

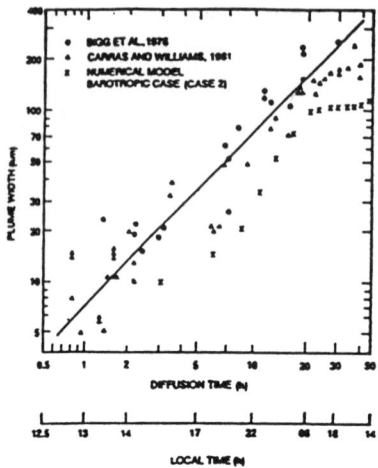

Figure 8b. Same as Fig. 8a except PBL shear is included.
(Reprinted from *Atmos. Environ.*, 1988, McNider, *et al.* with kind
permission from Elsevier Science Ltd.)

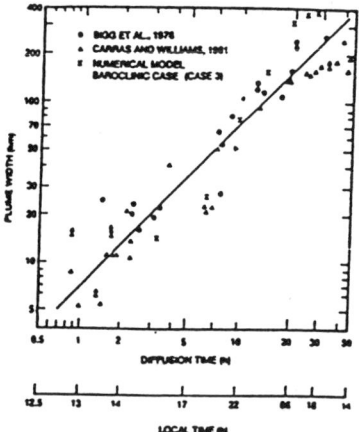

Figure 8c. Horizontal spread for short-term release (pseudo-puff) into
baroclinic flow. Includes PBL turbulence, PBL shear and geostrophic shear.
Local time applies to meteorological simulation results.
(Reprinted from *Atmos. Environ.*, 1988, McNider, *et al.* with kind
permission from Elsevier Science Ltd.)

In order to study the spatial and diurnal behavior of the dispersion rate, the results derived from the particle model are characterized by using the lateral standard deviation statistic, σ_y. Plots of σ_y are made versus both time and distance. The classical Pasquill-Gifford-Turner (PGT) (Turner (1971)) curves are superimposed onto the distance plots so that the results of horizontal dispersion could be compared with the standard dispersion rates used in operational models. The PGT curves are provided simply for reference. We realize these curves were never intended to be used for long transport times. The σ_y statistic represents the cross-wind spread of the plume and is computed as the standard deviation of particle position from the mean position of all particles. The exact behavior with distance is dependent upon release time. The σ_y statistics from particle model represent time-dependent information and are not necessarily equivalent to the original PGT curves which are based on steady-state conditions.

3.1.   Lateral dispersion statistics from the model experiments

Figure 9 shows a series of graphs for experiments 1-4 depicting horizontal dispersion with distance for different latitudes. When releases are made in the lower latitudes the results follow the more unstable curves at the beginning of the trajectory. These curves correspond to a daytime dispersion rate. This is of course due to more radiative heating at lower latitudes in the same season. Horizontal dispersion as a function of time for the same cases is shown in Fig. 10. The dispersion curves before sunset follow almost the same slope in each case, but the dispersion rate changes after sunset, especially for releases made in higher latitudes. During the daytime, horizontal dispersion is primarily controlled by the turbulent wind fluctuations. While there may be modest shear present in the daytime boundary layer, because particles move vertically through this shear, the resulting horizontal spread due to the shear is virtually canceled. However, during the night, vertical turbulence decreases, and vertical wind shear dominates the horizontal dispersion. Vertical shear develops between the new nocturnal boundary-layer (lower layer) and the old daytime boundary-layer (residual layer). At sunset due to decreased friction, wind directions in both the lower and upper layers turn toward lower pressure. Through the night, however, the wind direction in the upper layer gradually turns back toward the higher pressure as the wind field goes through an inertial oscillation in the process of adjusting to the geostrophic values of the frictionless state.

The magnitude of vertical shear is controlled by the amplitude and duration of the inertial oscillation. The duration of the oscillation is controlled by the length of night or more specifically by the length of time between the initial establishment of the nocturnal boundary layer and the complete growth of a new convective boundary layer the next morning.

In the night time shear layer, particles are in effect frozen at the vertical level they occupied at sunset. Thus, each level's particles move at the local wind direction and speed. Thus, at night, because of a lack of vertical movement, particles separate substantially in the horizontal due to shear.

As can be seen in Fig. 10, shear effect dominates the dispersion process. This is shown even more dramatically in Fig. 11 for cases 5-8 in which latitude is the same as in cases 1-4 but larger geostrophic wind speeds are imposed. The daytime dispersion rate does not increase significantly, but the nocturnal dispersion rate increases dramatically because of the larger vertical shear resulting from the higher wind speed.

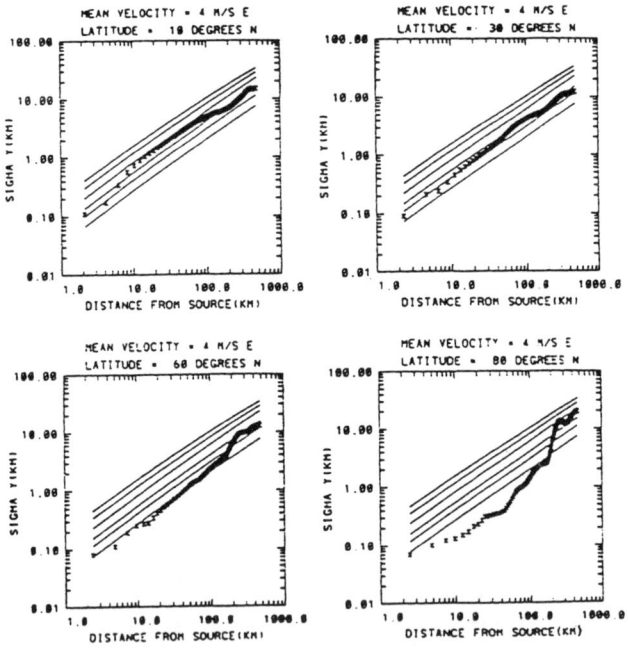

Figure 9. Horizontal dispersion with distance and reference Pasquill-Gifford-Turner curves.
(Reprinted from *Atmos. Environ.*, 1993, McNider, *et al.* with kind permission from Elsevier Science Ltd.)

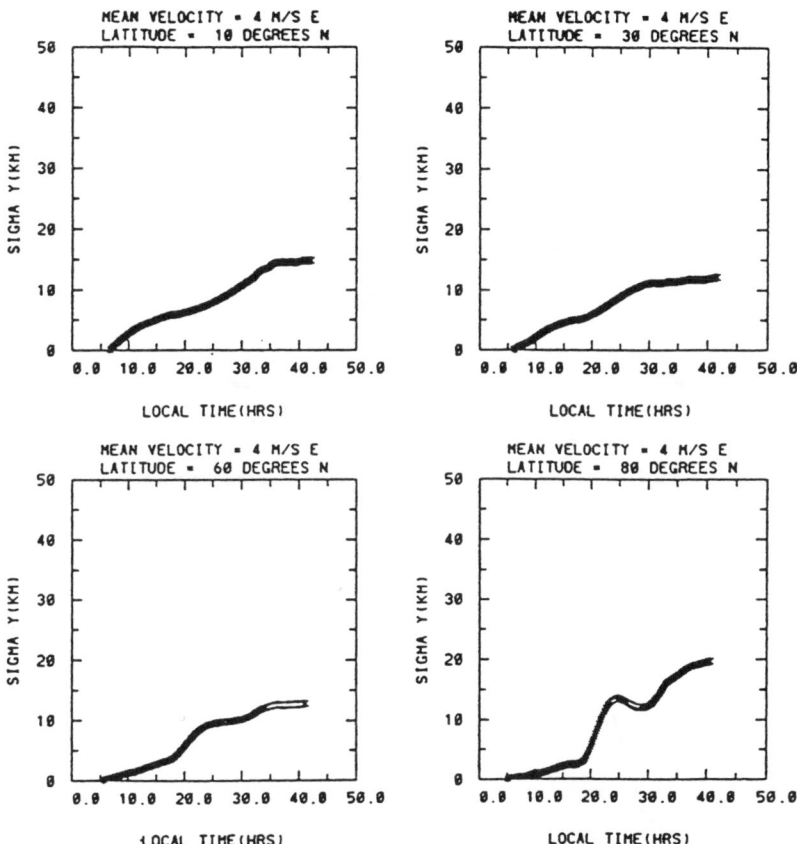

Figure 10. Horizontal dispersion with time. The ordinate gives time in hours in local time. For times greater than 24 subtract 24 h to give local time.
(Reprinted from *Atmos. Environ.*, 1993, McNider, *et al.* with kind permission from Elsevier Science Ltd.)

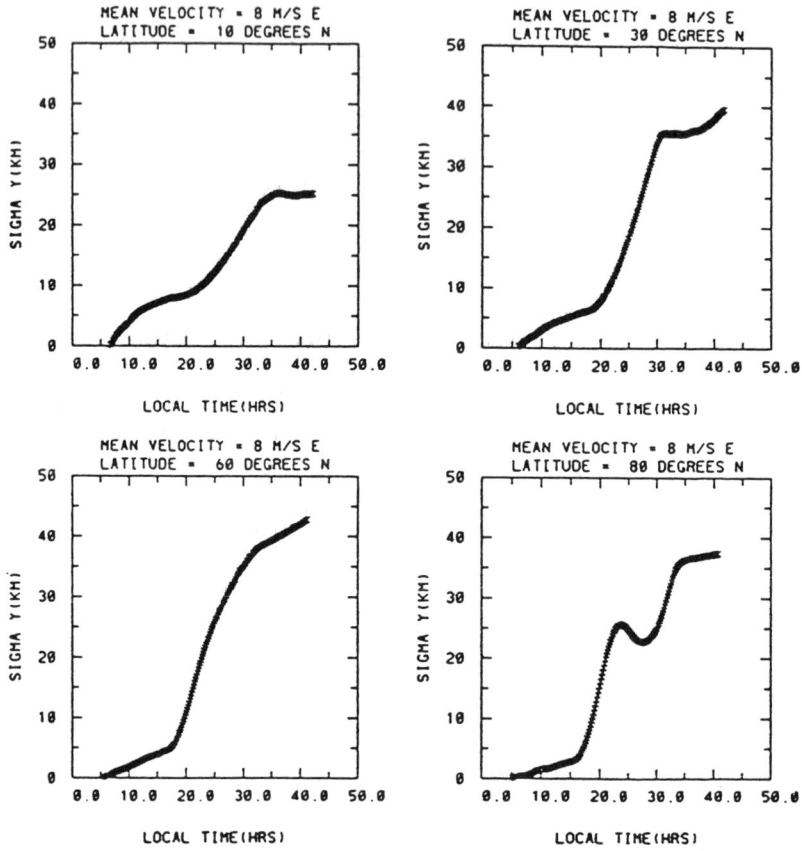

Figure 11. Horizontal dispersion with time. The ordinate gives time in hours
in local time. For times greater than 24 subtract 24 h to give local time.
(Reprinted from *Atmos. Environ.*, 1993, McNider, *et al.* with kind
permission from Elsevier Science Ltd.)

3.2   Observations of Boundary Layer Oscillations

Recently radar profiling technology has provided continuous measurements of the nocturnal boundary layer that captures details and climatologies of the inertial oscillation. Recently Gupta *et al.* (1997) carried out an analysis of the energy spectra from radar profilers. The following summarizes the study by Gupta *et al.* (1997).

In an air pollution field study Southern Oxidants Study (SOS) during June-July 1995 five 915 MHz boundary-layer radar profilers were operated in Nashville, Tennessee area. The profilers provide horizontal and vertical velocity components at a series of sample heights in the atmosphere. For this program the profilers had vertical gate spacing of 60 and 105 m and extended from a lowest gate at 120 m to the highest gate at 3880 m. See Ecklund *et al.* (1988), Neff *et al.* (1991), and White *et al.* (1991) for a more complete description of the operating characteristics of the radar profilers. The hourly consensus-averaged wind data (Fischler and Bolles (1981)) are used for the spectral analysis.

The time series of horizontal wind components were treated in several ways before the spectra were calculated. First, missing data points were replaced by linearly interpolating between adjacent good points. Then, the time series was detrended by subtracting the straight line best-fit from the time series, leaving a modified time series (Stull (1993)). For the modified time series the nonnormalized spectral density *(S)*, using the Tukey-Hanning spectral window, was computed with an algorithm described by IMSL, Inc. (1987). Next, $nS(n)$ vs. *log p* were plotted, where n is frequency in cycle per hour and p $(=1/n)$ is time period in hours. This procedure was followed for each site and altitude.

Figures 12 (a and b) show the relative horizontal wind components (u and v) spectra at all the range-gates for one of the profilers. In a more traditional depiction, Figures 12b and d show the spectra of horizontal wind components at the lowest range gate (120 m) and the range gate at 2.05 km for Dickson, TN. Figures 13 and 14 are same as Figure 12 except for Hendersonville, TN, and Youth Inc., TN, respectively. The important features in this display are the spectral energy peak near the diurnal period, most pronounced at low levels, and the larger broader peak at synoptic temporal period (150-200 hours). Note, as might be expected, that the synoptic peak is most pronounced at higher levels and decays toward the surface while the diurnal peak decays with height. In fact, the amplitude of the spectral peak is actually greatest just above the surface. As will be seen later, this differential spectral energy reflects variability of low level wind which is an important feature in maintaining plume growth rates. At this latitude the inertial period ($2p/f$), where f is the coriolis parameter, is about 20 hours which is close to the diurnal period. The peak in the spectra is a combination of both periods, and we will refer to this as the diurnal/inertial peak. This depiction shows the importance of the inertial oscillation in affecting the wind climatolgy in the boundary layer.

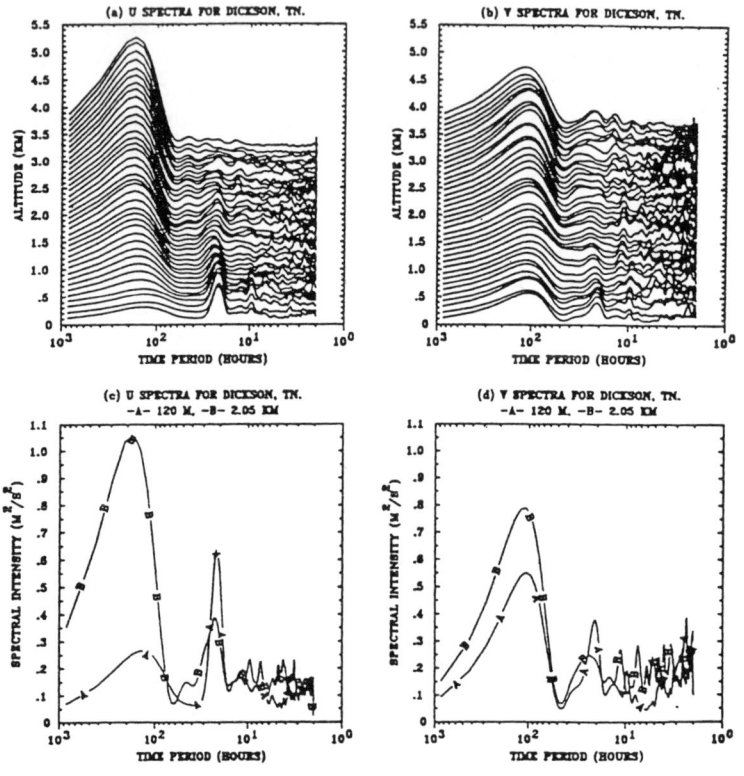

Figure 12 a and b. The relative energy spectra of horizontal wind components (U and V) for Dickson, TN. Constants are added to the energy spectra amplitude at successive heights to separate them on the graph. The corresponding height of a spectrum can be read from the left axis. (c) and (d) Horizontal wind components spectra at the lowest (120m) and 2.05 km range gates for Dickson, TN.
(Reprinted from *J. Appl. Meteor.*, 1997, Gupta, *et al.*)

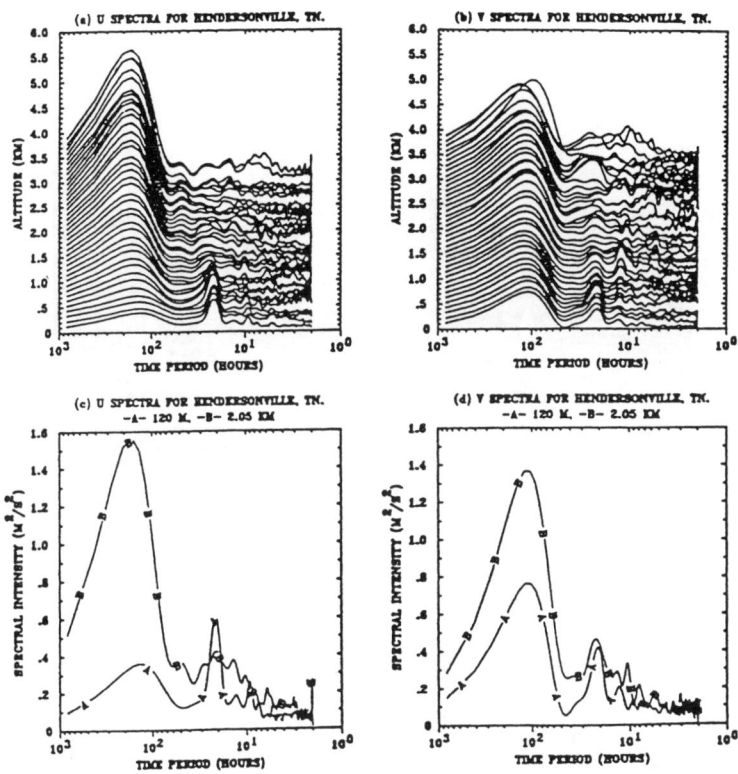

Figure 13. Same as Fig. 12 except for Hendersonville, TN.
(Reprinted from *J. Appl. Meteor.*, 1997, Gupta, *et al.*)

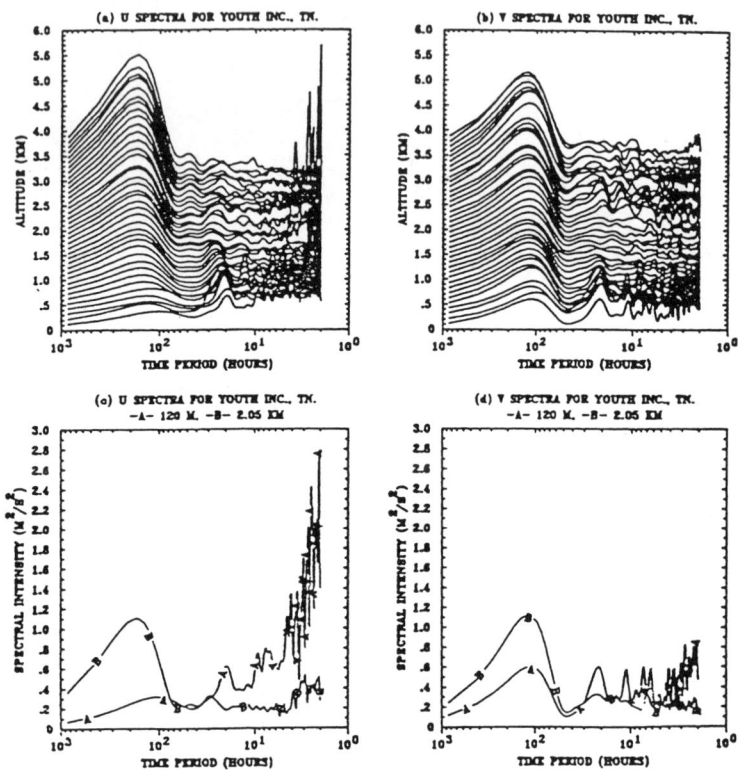

Figure 14. Same as Fig. 12 except for Youth Inc., TN.
(Reprinted from *J. Appl. Meteor.*, 1997, Gupta, *et al.*)

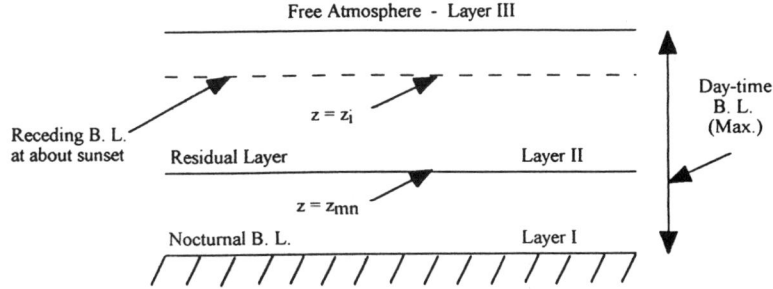

Figure 15. Layered structure of the PBL.

3.3   Parametrization of Inertial Oscillation in analytical form

Most operational air quality models, which are analytical rather than numerical, do not properly treat the role of inertial oscillations in the horizontal dispersion process. Analytical descriptions of boundary layer processes are needed to parameterize the effects of shear in these models. The first purpose here is to develop an analytical model than can be incorporated in operational models. Second, we use analytical solutions to show the characteristics of the nocturnal boundary layer that are controlled by external parameters. The dependency is best described analytically rather than through a series of sensitivity runs of a numerical model. The discussion, herebelow, is based on the analysis developed in Singh *et al.* (1993).

The governing equations in the PBL employing first-order turbulence closure are:

$$\frac{\partial u}{\partial t} = f(v - v_g) + \frac{\partial}{\partial v}\left[K_m \frac{\partial u}{\partial z}\right]$$

$$\frac{\partial v}{\partial t} = -f(u - u_g) + \frac{\partial}{\partial z}\left[K_m \frac{\partial v}{\partial z}\right]$$

(4)

where u is the East-West velocity component, v is the North-South velocity component, f is the Coriolis parameter, and K_m is the vertical turbulent exchange coefficient for momentum.

Since we are interested in the diurnal variation of the boundary layer, which greatly depends on forcing by the sun, we assume that

$$K_m = K_m(t) \ and \ \xi = z \, / \, \sqrt{K_m(t)} \qquad (5)$$

The solution in the PBL can now be written in the following form (see Singh *et al.* (1993)):

$$u = -v_g \exp(-\xi \, \sqrt{f/2}) \, \sin \xi \, \sqrt{f/2}$$

and

$$v = -v_g \left[1 - \exp(-\xi \, \sqrt{f/2}) \, \cos \xi \, \sqrt{f/2}\right] \qquad (6)$$

We have assumed that the geostrophic wind is directed along the y-axis, i.e. $u_g = 0$. This is basically an Ekman solution with time dependence included through $K_m(t)$.

The boundary layer height z_H can be determined from the exponentially decaying solution (eqn. 6) by putting, for example,.

$$\xi \, \sqrt{f/2} \cong \pi, \text{i.e.,} \quad z_H \sim \pi \, \sqrt{2K_m(t)/f} \qquad (7)$$

Relation (7) describes the structure of the PBL realistically in the sense that (7) predicts the depth of the layer to be comparatively larger during the day and lesser during the night. Solution (6) and relations (5) and (7) lead to the following three layers examined by Thorpe and Guymer (1977) (Figure 15).

Above the daytime PBL we have the free atmosphere (Layer III - frictionless) in which we have assumed geostrophic balance. The daytime PBL is composed of layers I and II. Layer I corresponds to the nocturnal boundary layer where the Ekman solution dominates. Maximum boundary layer height occurs at about noon time corresponding to maximum eddy viscosity. As boundary layer height recedes after midday, the atmosphere above the PBL will be in geostrophic balance, and the wind profile in the PBL will be governed by the Ekman solution.

At about sunset we assume that the net ground heat-flux changes sign and gives rise to an inertial oscillation (IO) in the atmospheric layer just above the inversion as the boundary layer recedes (Figure 15).

3.3.1   Formulation of an Initial-Value Problem for the Development of IOs

In the nocturnal boundary layer we assume that IOs are produced as a result of the horizontal momentum released because of the deviation from the geostrophic wind in the declining boundary layer. A characteristic feature of mixing height is that it increases after sunrise, attains a maximum sometime in the afternoon and remains stationary for a while. Afterwards it decreases slowly until about sunset when it drops sharply so that within an hour thereafter it attains its minimum value and remains almost stationary until daybreak (Singh *et al.* (1993)). In view of (eqn. 7), a realistic eddy-diffusivity coefficient profile is proposed below which takes into account the diurnal variation of mixing height in the real atmosphere.

$$
K_m = \begin{cases}
K(1-\omega), & t_{mn} \le t \le 24 \; or \; 0 \le t \le t_{mo} \\
K\left[1-\omega\cos\pi\left(\dfrac{t-t_{mo}}{t_{mx}-t_{mo}}\right)\right], & t_{mo} \le t \le t_{mx} \\
K(1+\omega), & t_{mx} \le t \le \tau \\
K\left[1-\omega\cos\pi\left(\dfrac{t-t_{mn}}{\tau-t_{mn}}\right)\right], & \tau \le t \le t_{mn}
\end{cases}
$$

$$(8)$$

where: t_{mo} = the time of sunrise, t_{mn} = time at which minimum mixing height is first attained, t_{mx} = time at which maximum mixing height is first attained, τ = time that the maximum mixing height persists ($\sim$ time of sunset, t_{ss}).

3.3.2   Solution for the residual layer (Layer II)

If we assume that the surface heat-flux reversal is felt in the upper layers of the boundary layer, say after one hour from sunset, we can expect IOs to be triggered in the upper layer at the same time. We can, accordingly, reformulate the problem for the generation of IOs for $z_{mn} \le z \le z_i$ during the period $t_i \le t \le t_{mo}$, where t_i represents the time of initiation of an IO and $t_i = t_{ss} + \tau'$.

$$\frac{\partial u}{\partial t} = f(v - v_g)$$

and

$$(9)$$

$$\frac{\partial v}{\partial t} = -fu$$

for $z_{mn} \leq z \leq z_i$ and $t_i \leq t \leq t_{mo}$ with the initial condition

$$\bar{v} - v_g \bar{j} = F(z)\bar{i} + G(z)\bar{j} \text{ at } t = t_i \tag{9a}$$

then the solution of (eqn. 9) for IOs will be

$$u = F(z) \cos f(t - t_i) + G(z) \sin f(t - t_i)$$
$$v - v_g = G(z) \cos f(t - t_i) - F(z) \sin f(t - t_i) \tag{10}$$

3.3.3   Solution for the nocturnal boundary layer (Layer I)

In the nocturnal boundary layer the wind profile is assumed to satisfy the following equations and boundary conditions:

$$\frac{\partial u}{\partial t} = f(v - v_g) + K_m \frac{\partial^2 u}{\partial z^2}$$

and $\tag{11}$

$$\frac{\partial v}{\partial t} = -fu + K_m \frac{\partial^2 v}{\partial z^2}$$

for $0 \leq z \leq z_{mn}$ and $t_i \leq t \leq t_{mo}$. K_m is determined from (8). We now impose the boundary conditions:

$$\text{(i) at } z = 0, u = v = 0$$
$$\text{(ii) at } z = z_{mn}, (t_i \leq t \leq t_{mo})$$

$$u = F(z_{mn}) \cos f(t - t_i) + G(z_{mn}) \sin f(t - t_i)$$

and

$$v - v_g = G(z_{mn}) \cos f(t - t_i) - F(z_{mn}) \sin f(t - t_i) \tag{12}$$

The nocturnal boundary layer solution satisfying (11) and b·cs (12) is

$$u = -v_g e^{-\pi z/z_{mn}} \cdot \sin \pi z/z_{mn} + \frac{z}{z_{mn}}[F(z_{mn}) \cos f(t - t_i) + G(z_{mn}) \sin f(t - t_i)]$$

and $\tag{13}$

$$v - v_g = -v_g e^{-\pi z/z_{mn}} \cdot \cos \pi z/z_{mn} + \frac{z}{z_{mn}}[G(z_{mn}) \cos f(t - t_i) + G(z_{mn}) \sin f(t - t_i)]$$

The most sensitive input is the initial wind profile (at the time when IOs are triggered). If we use more realistic initial data, e.g., observed wind at various heights below the free atmosphere at about the time of sunset $(t = t_i)$ solutions (10) and (13) are expected to give better approximations to the behavior of the real atmosphere in the nocturnal boundary layer. As an illustration, suppose we take linear structure functions:

$$F(z) = V \sin a \cdot \frac{z_i - z}{z_i - z_{mn}}, \ G(z) = V \cos a \cdot \frac{z_i - z}{z_i - z_{mn}} \tag{14}$$

where V and a are estimated from the numerical mesoscale model (McNider $et\ al.$ 1988). Figures 16 and 17 show the wind profiles produced by the analytical model. They demonstrate that the model's wind profile matches that of the numerical model (Figs. 18 and 19). The use of the mesoscale model is one way to specify the initial structure functions. Alternatively empirical boundary-layer similarity structure functions could be used.

Comparison of the model results with those of a mesoscale numerical model (McNider $et\ al.$ 1988) is also satisfactory (Figs. 16-19). We have obtained a closed-form solution of the governing Equations (9) (valid in the IO regime) and Equation (11) (valid in the nocturnal boundary-layer). The most sensitive input is the wind profile at the time when IOs are triggered. If we take more realistic initial data, such as the observed wind data at various heights below the free atmosphere at about the time of sunset $(t = t_i)$, solutions give better approximations to the real atmosphere in the nocturnal boundary-layer. There is still some uncertainty on the timing of the initiation of the IOs as a function of latitude. Additional studies and comparision with observation need to be carried out to clarify this point.

It may be remarked here that if the geostrophic wind v_g is strong, the NBL (Layer I) remains turbulently connected to Layer II (Fig. 15) and the analysis developed above will have to be modified.

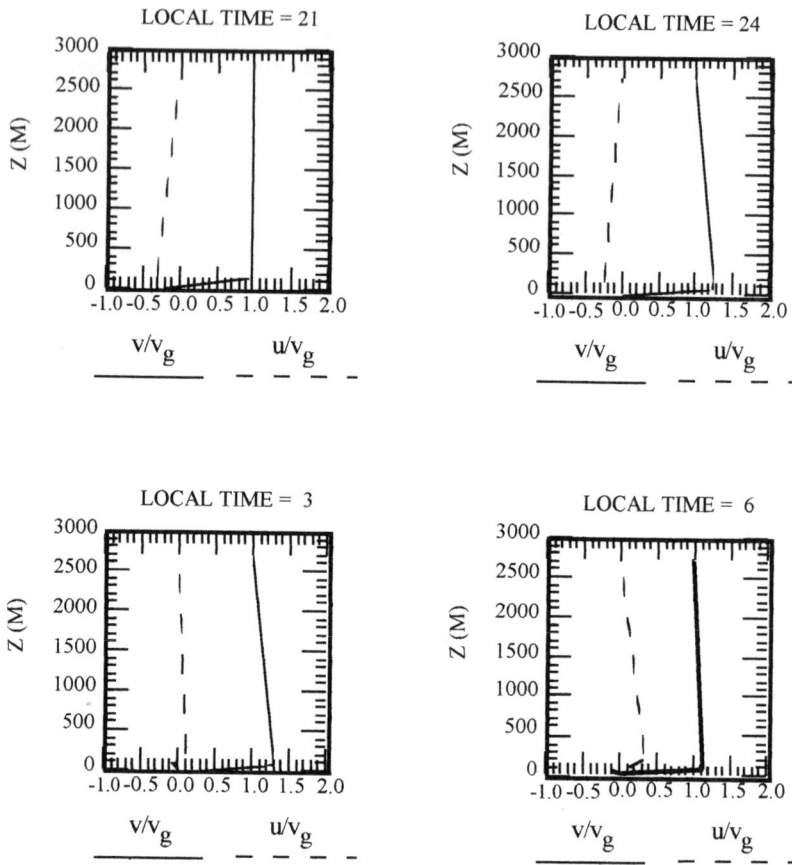

Figure 16. Wind profile vs. height at $\phi = 30°N$, $v_g = 8ms^{-1}$ (analytical model using solutions (10) and (13)).
(Reprinted from *Bound. Layer Meteor.*, 1993, Singh, *et al.*)

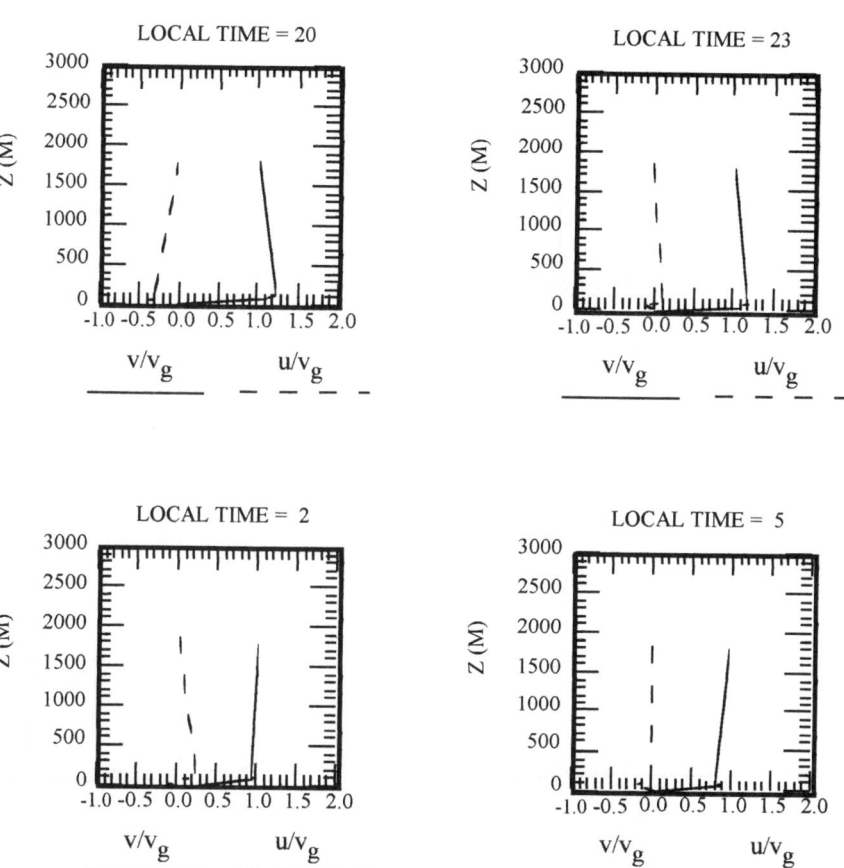

Figure 17. Wind profile vs. height at $\phi = 60°N, v_g = 8ms^{-1}$ (analytical model using solutions (10) and (13))
(Reprinted from *Bound. Layer Meteor.*, 1993, Singh, *et al.*)

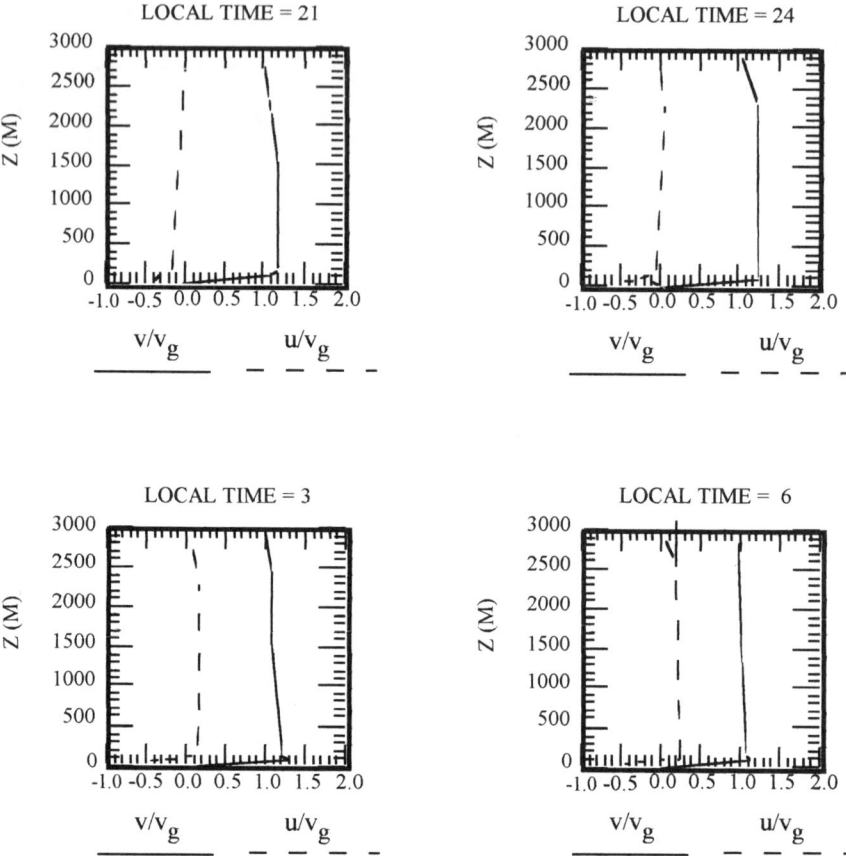

Figure 18. Wind profile vs. height at $\phi = 30°N, v_g = 8ms^{-1}$ (numerical model).
(Reprinted from *Bound. Layer Meteor.*, 1993, Singh, *et al.*)

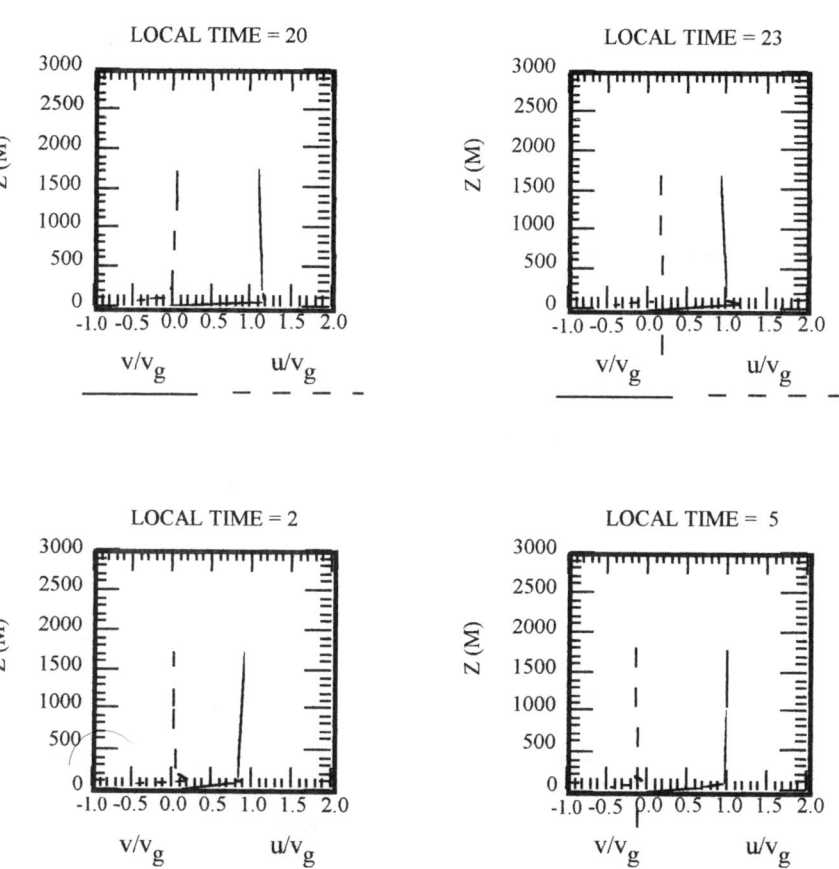

Figure 19. Wind profile vs. height at $\phi = 60° N, v_g = 8ms^{-1}$ (numerical model).
(Reprinted from *Bound. Layer Meteor.*, 1993, Singh, *et al.*)

4   References

Blackadar, A.K., 1957: Boundary-Layer Wind Maxima and Their Significance for the Growth of the Nocturnal Inversion. *Bull. Amer. Meteorol. Soc.*, **38**, 283-290.

Blackadar, A. K., 1979: High resolution models of the planetary boundary layer. Advances in Environmental and Scientific Engineering, Vol. I. Gordon and Breach, New York.

Businger, J. A., 1973: Turbulent transfer in the atmospheric surface layer. Workshop on Micrometeorology. D.A. Haugen, ed., Amer. Met. Soc., Boston, MA, pp67

Carras, J.N.and D.J. Williams, 1981:The long-range dispersion of plume from an isolated point source. *Atmos. Environ.*, **15**, 2205-2217.

Doedel, E. J., and J. P. Kernevéz, 1986: AUTO: Software for continuation and bifurcation problems in ordinary differential equations. User Manual "Auto 86", California Institute of Technology, Applied Mathematics Program Report, 226 pages.

Ecklund, W.L., D.A. Carter, and B.B. Balsley, 1988: A UHF wind profiler for the boundary layer: Brief description and initial results. *J.Atmos.Oceanic Tech.* **5**,432-441

England, D. E, and R. T. McNider, 1995: Stability functions based upon shear functions. *Bound. Layer Meteor.*, **74**, 113-130.

Fischler, M.A. and R.C. Bolles, 1981: Random sample consensus: A paradigm for model fitting with applications to image analysis and automated cartography. *Commun. Assoc. Mach.* **24**, 381-395

Garratt, J.R. and R.A. Brost, 1981: Radiative Cooling effects within and above the Nocturnal Boundary Layer. *J.Atmos. Sci.*, **38**, 2730-2746.

Gifford, F.A., 1983: Atmospheric diffusion in the mesoscale range: the evidence of recent plume width experiments. Preprint Volume, Sixth Symposium on Turbulence and Diffusion. Boston, March, Amer. Met. Soc., Boston, MA, 300-304.

Gifford, F.A., 1985: Atmospheric diffusion in the range 20-2000 Kilometers. In Air Pollution Modeling and its Application,V (edited by DeWispellaere C., Schiermeir F.A. and Gillani N.W.), 247-252. Plenum Press, New York.

Gupta, S., R.T.McNider, M.Trainer, R.J.Zamora, K.Knupp and M.P.Singh,1997: Nocturnal Wind Structure and Plume Growth rates due to Inertial Oscillaions. To appear in J. *Appl. Met.*

Husar, R.B., D.E. Patterson, J.D. Husar, N.V. Gillani and W.E. Wilson, 1978: Sulfur Budget of a Power Plant Plume. *Atmos. Environ.*, **12**, 549-568.

IMSL User's Manual Stat/Library, 1987: IMSL Inc.,2500 Park West Tower One, 2500 City West Blvd., Houston, Texas 77042-3020.

Lorenz, E. N., 1963: Deterministic nonperiodic flow. *J. Atmos. Sci.*, 130-141.

Lorenz, E. N., 1982: Atmospheric predictability experiments with a large numerical model. *Tellus*, **39**, 505-513.

McNider, R. T., and R. A. Pielke, 1981: Diurnal boundary layer development over sloping terrain. *J. Atmos. Sci.*, **38**, 2198-2212.

McNider, R.T., M.D. Moran, and R.A. Pielke, 1988: Influence of diurnal and inertial boundary layer oscillations on long -range dispersion. *Atmos. Environ.*, **22**, 2445-2462.

McNider, R. T., M. P. Singh, and J. T. Lin, 1993: Diurnal Wind-Structure variations and dispersions of pollutants in the boundary layer. *Atmos. Environ.*, **27A**, 2199-2214.

McNider, R.T., D.E. England, M.J. Friedman and X. Shi, 1995: Predictability of the stable atmospheric boundary layer. *J.Atmos. Sci.*, **52**, 1602-1614.

McNider, R.T., M.P. Singh, M.J. Friedman and X. Shi, 1996: Oscillations and intermittency in the Nocturnal Boundary Layer. Tech.Rep. ESSL, UAH, Huntsville, AL.

Moran, M.D., R.T. McNider, and R.A. Pielke, 1987: Diurnal influences on mesoscale atmospheric dispersion. Preprint Volume, Third Conference on Mesoscale Processes, Vancouver, August. Amer. Met. Soc., Boston, MA. 220-221.

Neff, W.J. Jordan, J. Gaynor, D. Wolfe, W. Ecklund, D. Carter and K. Gage, 1991: The use of 915 Mhz wind profilers in complex terrain and regional air quality studies. Preprints, 7th joint conference on Applications of Air Pollution Meteorology, New Orleans, Amer. Meteor. Soc., J230-J233.

Pasquill, F., 1962: Some observed properties of the medium scale diffusion in the atmosphere. *Quart. J. Roy. Meteor. Soc.*, **88**, 70-79.

Revelle, Douglas O., 1993: Chaos and "bursting" in the planetary boundary layer. *J. Applied Meteor.*, **32**, 1169-1180.

Richardson, L.F., 1926: Atmospheric diffusion shown on a distance-neighbour graph. *Proc. R. Soc.* (Lond) **A110**, 709-737.

Seydel, R., 1988: From equilibrium to chaos: Practical bifurcation and stability analysis. Elsevier Science Publishing Co., Inc., 367 pp.

Shi, X., 1997: Numerical investigation of the stable nocturnal boundary layer. Ph.D. Dissertation, University of Alabama in Huntsville, Huntsville, Alabama 35899.

Singh, M. P., R, T. McNider, and J. T. Lin, 1993: An analytical study of diurnal wind structure variations in the boundary layer and the low-level nocturnal jet. *Bound. Layer Meteor.*, 397-423.

Staley, D. O., and G. M. Jurica, 1972: Effective atmospheric emissivity under clear skies. *J. Appl. Meteor.*, **11**, 349-356.

Stull, R.B., 1993: An Introduction to Boundary Layer Meteorology. Kluwer Academic Publishers, Dordrecht, pp. 313. IMSL, Inc 87.

Thorpe, A.J. and T.H. Guymer, 1977: The Nocturnal Jet. *Quart. J. Roy. Meteorol. Soc.*, **103**, 633-653.

Turner, D.B., 1971: Workbook of atmospheric dispersion estimates. U.S. Environmental Protection Agency, Office of Air Program, Ref. No. AP-26.

White, A.B., C.W. Fairall, and D. Wolfe, 1991: Use of 915 Mhz wind profiler data to describe the diurnal variability of the mixed layer. Preprints, 7th joint conference on Applications of Air Pollution Meteorology, New Orleans, Amer. Meteor. Soc., J161-J166.

5   Acknowledgements

The authors would like to thank William B. Norris for his review of the manuscript, Cindy Taylor for her help in manuscript preparation, and Devdutta S. Niyogi for his assistance with formatting and editing. This work was supported in part by the Division of Atmospheric Sciences, National Science Foundation under Grant ATM-9120321.

Chapter 3
Urban air pollution

M. P Singh[a], J. Shah[b], and Hasnah Bte Hassan[c]

[a]*Earth System Science Laboratory, University of Alabama in Huntsville, AL 35899, USA*
[b]*Environment and Natural Resources Division, The World Bank, Washington D.C., 20433, USA*
[c]*Department of Mathematics, University of Brunei Darussalam*

Abstract

This chapter deals with urban air pollution. Health effects and economic impact are described. Summary of pollution assessment and the sources of pollution are outlined. Implications of dispersion and transport of pollutants aggravating the air quality are presented. Recent approaches in air quality management and related action plan are discussed. An overview of the current status of air pollution in twenty megacities based on a recent WHO/UNEP publication is contained in the appendix.

1 Introduction and Overview

Air pollution can be described as an atmospheric cocktail of dust and fine particulate matter along with ozone, nitrous oxide, sulfur dioxide and other gases, which gets inhaled and trapped in the lungs of pedestrians, motorists and urban dwellers at work and play. While this cocktail may not be inebriating, it is now widely regarded as a major health hazard. Unhealthy and sometimes lethal concentrations of pollutants have been released into the air for decades. The well-known London episode of December 1952 is often cited as a landmark in the understanding of air pollution's effect on human health. The increasing use of petroleum products for vehicles in urban areas during the last few decades has characterized a new form of urban pollution called photo-chemical smog. This was first encountered in 1945 in Los Angeles. Ozone, caused by toxins from vehicle exhaust reacting with air in bright sunlight, is thought to sensitize the body to the effects of irritants such as pollen and house dust.

According to a 1994 World Bank report, half the world's population will reside in urban areas within a decade. By 2015, the world will have 27 megacities, 17 of which will be in Asia. The urban population in developing countries is growing at an annual rate of 3.8 percent, and will increase from 1.4 billion people in 1990 to 3.6 billion in 2020. This swelling of urban populations, accompanied by industrialization and a growth in the number of motor vehicles, has resulted in a sharp rise in urban air pollution. More than a billion urban residents in developing countries breathe air that is harmful to their health. The health impacts of this air pollution, while not always calculated, can cost several billions of dollars. The poor and homeless living by the road-side are the worst affected.

Motor vehicle usage has increased tremendously throughout the world. In 1950, there were about 53 million cars on the world's roads; only four decades later with an average growth of 9.5 million automobiles per year, the global automobile fleet is over 430 million, more than an eight-fold increase. In addition to cars, there is a growth in the use of two-wheeled vehicles of which there are approximately 100 million. Adding these to the automobiles, the global motor fleet numbers 675 million (Mage & Zali 1992).

It is estimated that more than 200 million tons (MT) of carbon monoxide (CO), nearly 146 MT of sulfur dioxide (SO_2) and 53 MT of nitrogen oxides (NO_x) are discharged annually into the earth's atmosphere. Asia is responsible for approximately 20 percent of global vehicle emissions of CO, NO_x and 10.5 percent of CO_2 (Midgely 1994). Much of this pollution is emitted in and around urban centers. If projected growth and demand for vehicular transport and electricity are met using current technologies and aggressive abatement measures are not adopted, air pollution in urban areas will intensify in the coming years. The carrying capacity of most megacities has already been exceeded and policies must be put in place to ensure that cities are better planned and less congested to make conditions livable.

Indoor air pollution and exposure to health damage must also be recognized and mitigated. Many people spend considerable time indoors where pollution sources and levels may be very different than what are normally measured and regulated. The common sources of indoor air pollution include cook-stoves, cigarette or tobacco smoke, various cleaners and deodorizers. Depending on the lifestyle of the urban population, the solutions to reduce exposure must include outdoor as well as indoor sources for a cost-effective control strategy.

The sharp rise in the number of big cities will also mean increased competition for resources and investments. At least $280 billion per year for the next 25 years is needed to finance infrastructure in Asia's cities. Programs and policies to control air pollution will compete with other social and economic programs for resources and investments.

2   Urban Air Pollution

Many of the world's major cities are facing a host of environment-related problems, not the least of which is deteriorating air quality. Exposure to air pollution is now an almost inescapable part of urban life throughout the world. World Health Organization (WHO) guidelines are being regularly exceeded in most urban centers, in some cases to a great extent. Although industrialized countries have made substantial progress in controlling air pollution problems over the last two decades, air quality is worsening, particularly in the larger cities of developing countries. Given the rate at which these cities are growing and the general absence of pollution control measures in most of them, the air pollution situation will probably worsen and the quality of life of city dwellers will continue to deteriorate. WHO has identified urban air pollution as a major environmental health problem deserving high prioritized action.

Of foremost concern are the health and well being of urban residents. The concentrations of ambient air pollutants which prevail in many urban areas are sufficiently high to cause increased mortality, morbidity, deficits in pulmonary function, and cardiovascular and neurobehavioural effects. For example, a vendor sitting in his shop on any major thoroughfare of Bangkok, Bombay, or Jakarta will have by the day's end breathed in toxic chemicals, colloids and gases equivalent to more than nine cigarettes. In Bangkok alone, 45 percent of the traffic control policemen were afflicted by lung diseases of varying degrees. In the same city, average lead levels in blood are so high that 35 micrograms of lead per deciliter of blood is considered acceptable, whereas the USEPA recommends 25 microgram per deciliter as an acceptable level (World Watch, Washington, DC, 1994). Indoor sources of air pollution, such as cooking fires and smoking tobacco, add to the concentrations and may result in even more severe exposures for people in their homes.

In addition to affecting health, air pollution also causes serious damage to vegetation, material resource such as buildings and works of art. Pollution impairs visibility which dissuades tourism. Finally, urban agglomerations are also the major sources of regional and global atmospheric pollution and emissions of greenhouse gases.

3   Health Effects And Economic Impact

The health effects of air pollution are widely felt and recognized. In industrialized countries, there are well established policy and technical measures to control air pollution. In developing countries, people often accept environmental degradation as a price they must pay for economic prosperity. This attitude has been changing since the 1992 Rio Summit. People are beginning to understand that prosperity and a healthy environment are not mutually exclusive. Many solutions are simple and economical in the short and long term.

Increased mortality, morbidity, and deficits in pulmonary function have long been associated with annual and daily average SO_2 and suspended particulate matter (SPM) concentrations above WHO guidelines. Such health effects vary according to the intensity and duration of exposure to various pollutants and the health status of the population exposed. The level of exposure may also be influenced by factors such as climate, nutrition, quality of housing, smoking, occupation, etc. Often at greater risk are those people suffering from respiratory and cardiopulmonary disease, pregnant women, infants, the elderly, hyper-responders and exercisers.

Motor vehicle emissions cause a wide range of health effects. Carbon monoxide (CO) has a high affinity for hemoglobin and is able to displace oxygen, resulting in cardiovascular and/or neurobehavioural effects. Nitrogen dioxide (NO_2) and ozone often act synergistically and acute exposure can cause inflammatory and permeability responses, lung function decrements, increases in airway reactivity, and general respiratory symptoms. Lead is believed to cause blood enzyme changes, anemia, hyperactivity and other neurobehavioural effects. Table 1 (source: document provided by World Bank) provides a summary of the characteristics and health effects of SPM, ozone, lead, CO, NO_2, SO_2, and volatile organic compounds (VOC).

Aside from the most obvious impact on human and ecosystem health, and damage to physical infrastructure, air pollution has other social costs. Tangible social costs are calculated in terms of commercial productivity and tourism losses, while intangible costs relate to the loss in welfare, aesthetics, recreation, natural environment, and social productivity. Even though it may not be possible to quantify some of these losses, they must be accounted for in current and future planning and policy making.

Economic analysis of the health effects of air pollution is one way of assessing whether the cost of pollution control is economically justifiable. Due to the difficulty of assigning monetary value to various categories of air pollution damage, cost-benefit analysis is generally limited to valuing health benefits, and costs of various control options. By adapting methodologies developed in western countries (dose-response functions and epidemiological studies), and using the limited data available in most major cities, one can estimate air pollution exposure and its health impact. Estimates of morbidity and mortality costs are key in these calculations. According to a World Bank report, the economic cost of air pollution health damage is estimated to be US$1 billion per year in Bangkok, Jakarta, and other Asian cities.

Air pollution's economic costs are manifested in lower labor force productivity, deterioration of infrastructure, reduced attractiveness of the city to tourists and foreign and domestic investors. Consequently, slower economic growth, in turn, limits the resources available for the modernization of industry and production processes, further deepening the pollution problem. However, it is possible to delink increased economic activity from the trend of increasing pollution. In cities where effective environmental management (environmental strategy, regulations, institutional capacity, economic incentives and penalties) have been introduced, the forces of substitution, efficiency, and technological innovation have lowered pollution levels substantially and set the stage for sustainable patterns of development.

4   Summary Of Air Pollution Assessment

In order to assess the problems of urban air pollution in a global context, WHO and the United Nations Environment Program (UNEP) initiated a detailed study of air quality in 20 of the world's 24 megacities (UNEP/WHO 1992). For the purpose of this study, megacities were defined as urban agglomerations with current or projected populations of 10 million or more by the year 2000. Selected urban areas included cities in all parts of the world, including two in North America, four in Central and South America, one in Africa, eleven in Asia, and two in Europe. These are not necessarily the world's most polluted cities. The primary reasons for singling them out are common serious air pollution problems, large land areas and large population (20 megacities in 1990 had an estimated 234 million inhabitants). A review of the present day situation of air pollution in megacities can serve as a warning and a guide to rapidly growing urban areas when identifying the problems and adopting solutions.

The assessment began in November 1990, and was based on air quality data obtained by the WHO/UNEP Air Quality Monitoring Program, a component of the Global Environment Monitoring System (GEMS). In addition to information on air quality, data on pollution sources and other indicators were also collected from city authorities. The assessment ended in October 1991 at a meeting of government-designated experts who reviewed the results and recommendations. This summary of some of the important results of this meeting includes a number of conclusions regarding action by national as well as international agencies. The lack of reliable information on air pollution, its effects on population, sources of pollution, economic valuation of the damage, and policy and control options are some of the drawbacks. The international community, through participation in *Earthwatch*, must seek to fill these gaps.

4.1 Suspended Particulate Matter

Suspended particulate matter, the pollutant commonly known as SPM, includes many physically and chemically diverse substances. SPM exists in solid or aerosol state in ambient conditions. Particle size varies from sub microns to 100 microns. Particles that are less than 10 μm in diameter, known as PM_{10}, penetrate the upper respiratory tract and cause adverse health effects.

For western megacities, GEMS/Air data, together with national data, show a decline in SPM concentrations in line with the decline in emissions over the past few decades. This change is largely due to particulate control and changes in fuel use which have influenced both emissions and ambient concentrations. SPM emissions have fallen in some cities (New York and London) despite an increase in power generation from coal. Generating capacity has generally shifted out of the cities, and more efficient combustion techniques with particulate removal equipment have been introduced. Coal washing and the introduction of clean fuel have also helped to reduce emissions.

In developing megacities undergoing rapid industrialization and growth in vehicular traffic, the trend is very different. SPM is the major air pollution problem in most of the current (and future) megacities in developing countries. In many of these megacities, SPM levels are three to four times the WHO guidelines. Although national and local standards exist, they are not enforced. The economic benefits of industrialization, both to the cities in question and to the nations in general, often appear to out-weigh environmental concerns.

Several megacities are characterized by very high natural dust levels which influence SPM concentrations. Beijing, Delhi, and Cairo are examples where local soil characteristics and climate greatly influence urban SPM levels. In general, road vehicles, resuspension of road dust, fuel combustion for industry, energy, and cooking, and refuse burning are important sources of SPM. Transport emissions are largely uncontrolled, vehicles are poorly maintained and use sub-standard fuel that is often adulterated and manufactured without adequate controls. All of these factors result in high SPM concentrations.

4.2 Lead

Several risk assessment and other environmental studies have pointed out that human exposure to lead is one of the most serious urban environmental problems. Adverse health impacts, such as behavioral problems, reduced intelligence in children, and cardiovascular diseases in adults, occur without a known lower threshold. Exposure to lead from mobile sources is significant (often the main contributor) in those countries where tetra-ethyl lead is still a major fuel additive. Levels of tetra-ethyl lead in petrol have been reduced and even phased out in many industrialized countries; this is reflected in the ambient airborne lead concentrations in Tokyo, New York City, Los Angeles and London.

While other sources of lead exposures should not be ignored, reducing and ultimately completely removing lead from gasoline is a cost-effective measure to tackle lead exposure problems. Additionally, the availability of unleaded gasoline also facilitates the use of pollution-control devices that can control vehicular pollution and limit ozone.

During the UN Habitat Conference in June 1996, the World Bank called upon all countries to accelerate their efforts for a global phase-out of leaded gasoline. To this end, the World Bank has been supporting regional and country initiatives. This phase-out of leaded gasoline is independent of the use of catalytic converters. It is estimated that the United States saved more than US$10 for every investment dollar in eliminating leaded gasoline as a result of reduced health costs, savings on vehicle maintenance, and increased fuel efficiency. A recent governmental meeting on the "Elimination of Lead in Latin America and the Caribbean" launched new efforts to help those countries that are lagging behind in their lead phase-out plans.

4.3 Sulfur Dioxide

GEMS/Air data, together with data from national and local authorities, show that sulfur dioxide (SO_2) concentrations have decreased in line with emissions in many megacities. This is largely a result of changes in industrial and domestic fuel use. In some cities, such as London, New York City and Sao Paulo, clean air legislation has forced industry and domestic users to switch from high sulfur coal and oil to less directly polluting fuels such as electricity and natural gas. In cities such as Bombay, the change has been brought about largely by shifting textile and other industries to the outskirts of the city, and by increasing the use of natural gas.

On the other hand, several cities show a trend in increased annual SO_2 concentrations and emissions, largely associated with coal use.

Increasing emissions in Delhi are mainly a result of rapid industrialization. Beijing and Shanghai are almost totally reliant on coal for electricity generation and industrial and domestic fuel; therefore, annual average concentrations regularly exceed WHO guideline values. In Bangkok and Sao Paulo, a large proportion of SO_2 emissions are attributed to transport.

Despite reductions in annual average concentrations, urban SO_2 levels show marked seasonal and diurnal variation. Short-term guideline exceedances are not rare and often correspond with temperature inversions and cold winter conditions. These pollution 'episodes' may pose a significant risk to health, causing increases in mortality and respiratory illness. High pollution episodes may involve a number of pollutants (eg. SO_2, NO_2, CO) that act synergistically, intensifying health effects.

4.4 Oxides Of Nitrogen

Fossil fuel combustion and, increasingly, motor vehicle transport are the principal sources of oxides of nitrogen (NO_x) in urban areas. Each category accounts for between a third to half of all NO_x emissions in most cities. Nitrogen dioxide (NO_2) is of greater potential risk to health than nitric oxide, as it is a respiratory irritant. Oxides of nitrogen are also important precursors of secondary photochemical pollutants, such as NO_2 and ozone.

Analysis of NO_2 data for megacities reveals that ambient concentrations are increasing in virtually all urban areas. The correlation between elevated NO_x concentrations and motor vehicles has been clearly demonstrated by wide-scale monitoring in London and Los Angeles. The most heavily congested areas and major transport corridors show the highest annual average concentrations. NO_2, like other urban air pollutants, is affected by local topography and meteorological conditions.

Effective measures to control oxides of nitrogen have to a large extent been countered by the phenomenal growth in motor vehicle transport. Even in U.S. cities and in Tokyo where strict emissions controls have been imposed on mobile and stationary sources, levels have only stabilized, not declined. Effective NO_x controls for stationary sources (e.g., catalysts) are already employed in many industrialized countries, and the distribution and refinement of this technology should be encouraged whenever possible.

In order to reduce NO_x emission, alternatives to fossil fuel are required. Promotion of public transport especially in cities and a switch to alternative fuels for personal transport will help. In the short term, improved transport and road planning would also help to reduce exposure.

4.5 Ozone

Data for ozone are less widely available than data on other pollutants. Cities with high motor vehicle populations and the necessary meteorological conditions, such as Los Angeles and Mexico City, tend to experience the worst ozone episodes. Ozone formation is a photochemical process, dependent upon solar radiation and emissions of NO_x and volatile organic compounds, chiefly from motor vehicles. National and WHO guidelines are exceeded in Los Angeles, Mexico City and Sao Paulo throughout the year.

Little is known about ozone concentrations in the cities of developing countries where few studies or monitoring programs exist. Considering the high increase in NO_x and VOC emissions (e.g. Delhi, Bombay) associated with increasing road traffic, and the presence of favorable meteorological conditions, it is likely that ozone will become increasingly important in the future. To a large extent, ozone control is dependent upon the control of NO_x and VOC.

4.6 Carbon Monoxide

It is widely accepted that the growth in carbon monoxide (CO) emissions has corresponded with the increasing number of automobiles and mileage in recent years. In many industrialized countries, the introduction of stricter mobile emissions-control legislation and more efficient engines have helped to reduce or stabilize ambient CO concentrations. In developing countries the trends are less obvious because of a lack of monitoring. Emissions inventories suggest that CO emissions from motor vehicles are increasing in parallel with NO_x emissions. This is reflected in studies on average concentrations and human exposure. Even where average urban population levels are declining, elevated curbside CO concentrations are a major concern, due to direct exposure of people at the road side.

The issues relating to CO control are similar to those for NO_x and ozone. Motor vehicles are the main source of CO pollution, especially in terms of direct human exposure. In the short term, the introduction of emissions control legislation similar to that enforced in the United States, Japan, and Europe may help to stabilize emissions.

5   Sources

Most aspects of urban life are influenced by energy production and consumption. Energy is required for cooking, heating and lighting, for motorized transport, and industrial processes. Fossil fuels meet the majority of these energy demands in cities throughout the world, either directly or via conversion to electrical energy. Growing urban populations and levels of industrialization inevitably lead to greater energy demand which is usually reflected in increasing pollutant emissions.

Combustion of fossil fuels for domestic heating, generating power, running motor vehicles, industrial processes, and disposing solid wastes by incineration are generally the principal sources of air pollutant emissions in urban areas. Historically, the most common air pollutants in urban environments include SO_2, nitrogen oxides (NO and NO_2, collectively termed NO_x), CO, ozone, SPM and lead.

Combustion is the principal man-made source of primary air pollutants. Fossil fuel combustion in stationary sources leads to the production of SO_2, NO_x, particulates - both primary particulates in the form of fly ash and soot; and secondary particulates, sulfate (SO_4), nitrate (NO_3), and aerosols (formed in the atmosphere following gas to particle conversion). Domestic solid fuel use, mainly coal and wood, also represents a significant source of these pollutants in the urban areas of developing countries. Petrol-fueled motor vehicles are the principal sources of NO_x, CO and lead, whereas diesel-fueled engines emit significant quantities of particulates and SO_2 in addition to NO_x.

Ozone, a photochemical oxidant and the main constituent of photochemical smog, is not emitted directly from combustion sources. It is formed in the lower atmosphere in the presence of sunlight from NO_x and VOC. VOC may be emitted from a variety of sources including road traffic, production and use of organic chemicals (e.g., solvents), transport and use of crude oil, use and distribution of natural gas and, to a lesser extent, from waste disposal sites and waste-water treatment plants. Cities in warmer, sunny locations with high traffic densities tend to be especially prone to the net formation of ozone and other photochemical oxidants from precursor emissions.

Although detailed emission inventories are not widely available on the basis of observed trends in national emissions and recent increases in vehicle registrations, it may be concluded that motor vehicles now constitute the main source of air pollutants in the majority of cities in industrialized countries. This is particularly true for CO, NO_x and SPM. In contrast, cities in developing countries have a greater variety of air pollution sources. The relative contributions of mobile and stationary sources to air pollutant emissions differ markedly between cities, and are dependent on the level of motorization and the level, density, and type of industry present. Cities in Latin America, for example, tend to have higher vehicle densities than those in other developing regions and, therefore, are likely to experience a higher contribution of air pollution from motor vehicles. Motor vehicle contributions are less important in cities with fewer vehicles (e.g., cities in Africa) and in cities located in temperate region that are dependent on coal or biomass fuels for space heating and other domestic purposes (e.g., cities in China and parts of eastern Europe). It is also worth noting that in many developing countries, vehicle fleets tend to be older and poorly maintained, a factor which raises the significance of motor vehicles as a pollutant source.

Presently, the world's vehicle fleet is concentrated in the high income economies. In 1988, Organization for Economic Cooperation and Development (OECD) countries alone accounted for 80 percent of the world's cars, 70 percent of trucks and buses and over 50 percent of two- or three-wheeled vehicles. Since 1950, the global vehicle fleet has grown tenfold and is expected to double from the present total of 630 million vehicles within the next 20-30 years. The rate of growth of the world's vehicle fleet is projected to out-pace that of both the total and urban population.

Much of the expected growth in vehicle numbers is likely to occur in developing countries and in eastern Europe. In contrast, much of the demand for motor vehicles in the developed countries will be for vehicle replacement. In the absence of the introduction of stringent control measures for traffic-related pollutants, the contribution of motor vehicles to the pollution load is likely to increase and it is expected that air quality will subsequently deteriorate in these regions.

Smoke emissions from diesel vehicles are a cause of considerable concern throughout the world. This source can account for up to 70 percent of SPM (smoke) emissions in some cities. Most emission estimates and projections show an increase in respirable particulates attributed to diesel vehicles, due to their increasing popularity combined with a lack of adequate control legislation. This is particularly worrisome due to the carcinogenic qualities of certain smoke constituents, e.g., polycyclic aromatic hydrocarbons.

In addition to the more common or "traditional" air pollutants, a large number of toxic and carcinogenic chemicals are increasingly being detected in urban air, albeit at low concentrations. Examples include selected heavy metals (e.g., beryllium, cadmium and mercury), trace organic (e.g., benzene, polychlorodibenzo-dioxins and furans, formaldehyde, vinylchloride and polyaromatic hydrocarbons (PAHs)), radionuclides (e.g., radon) and fibres (e.g., asbestos). Such chemicals are emitted from a range of sources including waste incinerators, sewage-treatment plants, industrial and manufacturing processes, solvent use (e.g., in dry-cleaning establishments), building materials and motor vehicles. Although such chemical emissions are lower than traditional pollutant emissions, they may pose significant risks to health because of their extreme toxicity or carcinogenic potential, or a combination of both. As the measurement of low concentrations of toxic chemical contaminants in air presents analytical difficulties, very little monitoring is currently conducted.

6   Dispersion And Transport

The earth's atmosphere is both a reservoir or receptacle for many of the by-products of human activity. It is a pathway by which these materials make their way from industrial and other sources to oceans, the biosphere, and humans. As in the case of the "greenhouse gases", long-lived pollutants injected into the atmosphere may sufficiently alter its properties and behavior to have significant adverse effects on human health and environmental well-being. Other pollutants such as acidifying emissions or toxic organic chemicals are delivered by the atmosphere and deposited on the sensitive parts of the ecosystem.

Urban air pollution problems are often aggravated by meteorological and topographical factors that concentrate pollutants in the city, and prevent proper dispersion and dilution. Many cities are surrounded by hills which may act as a downwind barrier, trapping pollution close to the city. Thermal inversions are a particular problem in temperate and cold climates. Under normal dispersion conditions, hot pollutant gases rise and come into contact with colder air masses with increasing altitude. However, under certain circumstances, the temperature may increase with altitude and an inversion layer forms a few tens or hundreds of meters above the ground. This inversion layer may then trap pollutants close to the emissions source and act as a heat cover prolonging the inversion. These conditions are of greatest concern when wind speeds are low. Isothermal conditions, when there is no change in temperature with altitude, may have a similar result. Another meteorological phenomenon which greatly influences air quality is the "urban heat island". The heat generated by a city causes the air to rise which draws in colder, and possibly more polluted, air from surrounding areas.

On a more local scale, buildings and other structures can have a significant effect on pollutant dispersion. Tall buildings on either side of a busy road prevent the dispersion of low level emissions, creating the "street canyon" effect.

It is apparent that the long-range transport of air pollution from megacities may have national and regional impacts. Oxides of nitrogen and sulfur in the "urban plume" may contribute to acid deposition at great distances from the city. Ozone concentrations are often elevated downwind of urban areas due to the time lag involved in photochemical processes and NO_x scavenging in polluted atmospheres.

It is now understood that air pollution is no longer just a local issue, mainly connected with urban/industrial point-sources. The regional dimensions of air pollution, associated with the long-range dispersion of acid precipitation and photochemical oxidants, and even global problems such as long-term climate change, are being more clearly understood. In addition to these "chronic" concerns that are shared by all countries, there is growing risk of acute manmade air pollution emergencies, the impacts of which transcend national boundaries. These range from recurrent, seasonal, smog alerts to major industrial accidents such as the lethal Bhopal and Chernobyl clouds, and even the burning of oil wells in Kuwait.

While environmental damage by trans-frontier air pollution in Europe and North America is now abundantly documented, there still is insufficient information on the situation in other parts of the world. A 1988 study by UNEP and the International Council of Scientific Unions (ICSU) has identified southern China and other areas of Southeast Asia, southwestern India, equatorial Africa including Nigeria, southeastern Brazil, and northern Venezuela, as being areas of probable sensitivity to soil and surface water acidification caused by long-range air pollution. The United Nations Economic Commission for Africa reports substantial levels of air pollution in the Zambian copper mining belt. The discharge of SO_2 and nitrogen oxides from this area may lead to the formation of acid rain which falls across the border into neighboring countries. Specific cases of the formation of smog from nickel mining activities have been observed in Zimbabwe and Angola. Extensive industrial activities in the Republic of South Africa are considered a possible source of transboundary air pollution in that sub-region.

The long-range transport of urban air pollutants results in impacts on a national and regional scale. Urban emissions of nitrogen and sulfur oxides may contribute to acid deposition at great distances from the source. Projected increases in urbanization, industrialization and a growth in the number of motor vehicles and the associated increase in emissions would suggest that megacities and other rapidly developing urban areas will become an important source of TRANSBOUNDARY pollution.

7   Air Quality Management

While an air quality management framework is in place in most cities in the industrialized world, in the majority of the developing world's megacities it has been given little consideration. Weak or non-existent environmental management systems are to blame for this state of neglect. Other reasons include: the lack of a single coordinating authority with sufficient resources; absence of analytical frameworks; no urban zoning and transport planning; inadequate monitoring; absence of appropriate ambient and source standards; and lack of enforcement. As economic development proceeds, it is becoming clear that strategies must be created to deal with urban air pollution initially from a policy and planning perspective but eventually through stricter standards, better monitoring, a higher level of public awareness, and finally, through source controls.

A comprehensive air quality management system (AQMS) includes both economic and air quality analyses. The steps include air quality monitoring, identification of source receptor relationships, identification of impacts and estimation of social and economic costs, establishment of air quality goals and time based targets, and enumeration of air pollution options. Based on this comprehensive assessment, the analytical framework for an air quality management strategy should be developed.

A recent review of the program has proposed that additional pollutants should be included in the GEMS/Air monitoring program, emissions inventories should be compiled, and additional countries should be encouraged to join in order to strengthen the world-wide coverage.

It is clear that air pollution in many of the world's megacities, as well as in other cities, is a major health and environmental concern, and should command high priority for action. National efforts to deal with this worsening problem are under way in many cities, but such efforts must be strengthened. Global and regional initiatives aimed at counteracting global warming and transboundary air pollution through measures such as cleaner production and energy conservation, will also result in reducing the emissions of many of the air pollutants of local concern. Additional strategies will, however, be needed to deal with air pollution within the cities. The WHO Commission on Health and Environment has identified the following as possible strategic elements for urban and transportation planning, particularly in fast-growing cities:

- provision of safe and convenient mass transport;
- control of emissions from vehicles (especially automobiles) and industries;
- a switch to cleaner fuels;

- emphasis on district heating with co-generating plants; and
- stringent controls on new industrial operations, and appropriate siting of power generating plants in relation to residential areas.

Combustion modification techniques such as low NO_X burners, fluidized bed combustion and flue gas desulfurization (FGD) have been employed in some countries to reduce NO_x and SO_x emissions. However, modern combustion technology is sometimes expensive and beyond the purchasing power of many developing countries.

Post-combustion control involves the removal of pollutants from flue gases and vehicle exhaust. There are many methods for removing SPM from flue gases, including gravity settling chambers, cyclones, spray chambers, bag filters and electrostatic precipitators. A combination of these techniques is usually employed on larger coal-fired boilers. Flue gas de-sulfurization is being increasingly adopted by many industrialized nations to meet emissions standards and to achieve internationally agreed targets. Catalytic reduction is at present the best available technique for reducing NO_x emissions. The introduction of simple chimneys and vents in domestic stoves and heaters has greatly improved indoor air quality in developing regions. The use of three-way catalytic converters (TWCs) to control motor vehicle pollutants has resulted in significant reductions of NO_x, CO and hydrocarbons emission from new vehicles.

The introduction of new manufacturing processes has led to significant reductions in industrial emissions. An obvious example is the use of low temperature hydrometallurgical techniques that reduce the SO_2 emissions associated with traditional metal smelting methods. Many countries have adopted energy conservation measures mainly on economic grounds, and have effectively limited energy consumption and demand and also improved generation and distribution efficiency. Responses to transboundary air pollution will inevitably impinge on a wide range of other environment and development activities. For example, nitrogen oxides emissions must be controlled locally because of their contribution to the formation of ground-level ozone and urban smog; regionally because of their share in acid precipitation; and globally because tropospheric ozone also acts as a greenhouse gas. International action in this field thus offers a number of multiple-benefits options to decision makers.

8   Summary

Clearly, air pollution in urban centers has worsened, leading to new levels of concern. The economic impacts of this pollution are now widely recognized and have been quantified as significant. The above observations can be summarized as follows.

a)   Pollutant concentrations in most of the megacities in developing countries, as measured by SPM, exceed the WHO guidelines. SPM levels will continue to rise with continued urbanization and growth in vehicular traffic.

b)   There are notable improvements with respect to certain pollutants in cities of industrialized countries. Even for those cities, oxides of nitrogen and ozone (e.g. Los Angeles) remain a problem despite stringent controls.

c) SO_2 concentrations have been declining in most cities. Cities that rely on coal as a fuel, such as in China, are an exception.

d)   Ambient lead levels are probably decreasing due to phasing out of leaded gasoline in many industrialized countries. However, low lead and unleaded gasoline are not yet available in many urban areas in the developing countries.

e)   There are still major difficulties in assessing the adverse health effects of urban air pollution due to lack of information on actual human exposure. This is particularly true in the cities of developing countries where socio-economic factors, climate conditions, etc. are very different from those areas where such research has been conducted. The scarcity of human health data has contributed to the lack of appreciation of the severity of air pollution problems in these cities.

f)   National capabilities for the assessment and management of air pollution problems are grossly inadequate in most countries.

9 Action Plan

In the context of above observations, the following action plan is recommended.

a)   Empower a single coordinating agency with a clear mandate and resources to carry out air quality management.

b)   Create and strengthen monitoring and assessment capabilities, and establish relationships and management alternatives in the developing countries. This can be done with knowledge and technology transfer through twinning arrangements with developed countries.

c)   Apply stringent emission control strategies to all new industrial developments.

d)   Implement innovative and progressive long-term development schemes and decentralization plans (e.g., better mass transit, combined heat and power stations, better treatment of waste, and urban planning).

e)   Vehicle Emissions Control

(i)   Address gross polluters: Existing smoke opacity regulations for diesel vehicles must be strictly enforced. The success of this action depends upon the routine maintenance and adjustment of engines.

(ii)   Clean vehicle emission standard: It is crucial to establish state-of-the-art vehicle emission standards for gasoline cars, diesel vehicles and motorcycles. Such a standard would be better enforced if the availability of lead free gasoline, at prices below that of leaded gasoline, could be assured.

(iii)   Fuel economy: Fuel economy standards, differential vehicle taxes, and bonuses for scrapping less fuel efficient vehicles are examples of the measures that should be adopted to promote greater fuel efficiency.

(iv)   Unleaded fuel: This is an important early action and a pre-requisite for clean vehicle standards. Differential fuel taxation may be used to promote the use of unleaded fuel, and leaded fuel should be progressively and rapidly phased out.

(v)   Diesel fuel: Fuel price surcharges based on the sulfur/heavy oil content of diesel should be introduced to promote the availability and use of cleaner fuel in trucks and buses.

(vi)   Alternative fuels: Incentives to switch to cleaner alternative fuels can be provided through differential fuel taxation or direct subsidy where environmental benefits would justify it.

(vii) Inspection and maintenance programs: These are needed for the enforcement of clean vehicle standards and are essential to ensure that the anticipated benefits of emission control strategies are not lost through poor maintenance or tampering with emission control devices.

f)   Improve public awareness via newspapers and television. Promote private sector initiatives (e.g. only accepting deliveries from vehicles and trucks that meet government standards by measurements at the company gate, adopt-a-street campaign, private pollution monitoring with displays at major intersections, etc.) to assist with the urban air quality improvement.

TABLE 1: POLLUTANT CHARACTERISTICS AND EFFECTS

Pollutant	Characteristics	Adverse Health Effects
Suspended Particulate Matter (SPM)	Major sources of SPM include soot and condensed vapors from combustion in vehicles; stationary combustors; open burning of agricultural and domestic wastes; windblown dust from devegetated areas; dust stirred up by vehicular traffic; and smelting and processing of non-metallic minerals. In Asian cities, on average over 60% of SPM is less than 10 microns in size (inhalable size range), frequently in the range of 0.6-1.0 microns, and 5-7 microns.	Exposure to high short-term levels of SPM has been linked to increases in illness and death from respiratory causes, especially when particulate include acid aerosols such as sulfate and nitrate particles, and especially in the presence of high levels of SO_2. Long-term exposure to high SPM levels results in increased susceptibility to respiratory illness, death from respiratory causes, and diminished lung function.
Ozone (O_3)	Ozone is a "secondary" pollutant, formed by chemical reactions in the atmosphere between oxides of nitrogen (NO_x), unburned hydrocarbons and other volatile compounds (VOC) in the presence of sunlight.	Short-term exposure to high ambient O_3 concentrations causes irritation and inflammation of the eyes and respiratory tract, increased mucous production and decrease in lung function. Long term exposure may lead to permanent lung damage.
Lead (Pb)	The major source of atmospheric lead is combustion of lead antiknock compounds in gasoline (approximately 90% of lead in urban air comes from this source).	Lead is a cumulative, systematic toxin with adverse effects on blood pressure regulation and the nervous system. At high levels, lead is an acute toxin, damaging the kidneys, nervous system, and brain.
Carbon Monoxide (CO)	Carbon monoxide is formed by the incomplete combustion of carbon-containing fuels. CO emissions come primarily from gasoline vehicles.	When inhaled, CO binds to hemoglobin in the blood to form carboxyhemoglobin. This reduces oxygen supply to the brain and heart, resulting in shortness of breath, increased blood pressure and headaches.
Nitrogen Dioxide (NO_2)	Nitrogen dioxide plays a key role in the photochemical production of ozone. It is a toxic gas, some of which is emitted directly as a result of combustion. However, most NO_x from combustion sources is emitted in the form of NO. This is subsequently oxidized to NO_2 by photochemical reactions. Key sources of NO_x emissions in Asian cities are gasoline and diesel vehicles, fuel combustion in power plants and industrial burners.	Short-term exposure to high levels of NO_2 causes lung edema and damage to lung cells, increasing susceptibility to bronchial infections. NO_2 also acts as a bronchi-constrictor, aggravating asthmatic conditions. Long-term exposure to low levels of NO_2 causes severe damage to lung tissues.
Sulfur Dioxide (SO_2)	SO_2 is produced by the combustion of sulfur containing fuels, smelting of sulfide ores, and in some Asian cities, emitted by petroleum refineries. In the atmosphere, SO_2 reacts with ozone to produce SO_3, which combines with water to form sulfuric acid.	SO_2 is absorbed primarily in the nasal system. The effects of exposure to high short-term concentrations include irritation of the respiratory tract and bronchitis. Longer exposure can cause severe respiratory illness.
Volatile Organic Compounds (VOC)	VOC emissions relate most directly to the negative health impacts from photochemical oxidants. A typical VOC would be benzene, a hydrocarbon constituent of gasoline.	Extended epidemiologic studies of benzene show a strong linkage to increased leukemia incidence.

Acknowledgement:

Authors have benefited from reports of many international institutions listed in the bibliography and would like to acknowledge their contribution. However, the findings, interpretations and conclusions expressed in this publication are entirely those of the authors and should not be attributed in any manner to any institution. The authors would also like to thank Mr. Devdutta S. Niyogi, Ms. Seema Singh, and Ms. Tanvi Nagpal for their assistance with formatting and editing of the manuscript.

10   Bibliography

Bradley, D. et al. (1991) *A review of environmental health impacts in developing country cities.* Urban Management and the Environment. World Bank-UNDP-UNCHS.

Butterwick, L., Harrison, R. and Merritt, Q. (1992) Handbook for urban air improvement 1991. Prepared for the Commission of the European Communities. Cambridge/London, Cambridge Decision Analysis/ Environmental Resources Limited.

Cairncross, F. (1991) *Costing the Earth.* The Economist Books/Business Books, London.

Cochran, L.S., Pielke, R.A. and Kovacs, E. (1992) Selected international receptor based air quality standards. *J. Air Waste Manage. Assoc., 42,* 1567-1572.

Economopoulos, A. P. (1993) Assessment of sources of air, water and land pollution. A guide to rapid source inventory techniques and their use in formulating environmental control strategies. Part I: Rapid inventory techniques in environmental pollution. Geneva, World Health Organization (WHO/PEP/GETNET/93.1-A).

Elangovan T., (1992), "Assessment of Transportation Growth in Asia and its effects on Energy Use, the Environment and Traffic Congestion: Case Study of Varanasi, India", International Institute of Energy Conservation (IIEC), The National Transportation Planning and Research Center (NATPAC).

Ericsson, G. and Camner, P. (1983) Health effects of sulfur oxides and particulate matter in ambient air. *Scand. J. Work, Environ. Health, 9,* suppl. 3.

Faiz. A., Weaver C. S., Sinha K., Carbajo J., (1996), The World Bank, "Air Pollution from Vehicles: Impact and Control Measures".

Freeman, A. Myrick (1982) *Air and Water Pollution Control.* John Wiley & Son, New York.

Govind, H. (1989) Recent Developments in Environmental Protection in India: Pollution Control. *Ambio* 18 (8), p. 429-433.

Hatzakis, A. et al. (1986) Short-term effects of air pollution on mortality in Athens. *Int. J. Epid.*, 15:73-81.

Hutcheson, R. and C. van Paassen (1990) *Diesel fuel quality into the next century*. Shell, Selected papers.

Krupnick, A.J. (1991) *The Valuation of Environmental Externalities: Guidance Document*. Resources for the Future, Washington D.C. (draft).

Larssen, S., et al. (1996) URBAIR. Urban air quality management strategy in Asia. Bombay, Jakarta, Manila and Kathmandu Valley specific reports. (World Bank, ASTEN Technical discussion papers).

Larssen, S., et al. (1996) URBAIR. Urban air quality management strategy in Asia; Guidebook (World Bank, ASTEN Technical Discussion paper).

Lave, L.B., E.S. Seskin (1977) *Air Pollution and Human Health*. The Johns Hopkins University Press, Baltimore/London.

Lowe, M.D. (1990) Cycling Into the Future. In: L.R. Brown *et al.*, *State of the World 1990*. W.W. Norton & Company, New York/London.

Mage D.T., Zali O. eds., (1992), WHO/PEP/92-4, "Motor Vehicle Air Pollution, Public Health Motor Vehicles, Issues and Options for Developing Countries".

Mayur, P. (1992) *Death Hovers over Bombay*. Urban Development Institute, Bombay.

Midgely, P. (1994) Urban transport in Asia. An operational agenda for the 1990s. World Bank technical paper number 224.

Mishan, E.J. (1988) *Cost Benefit Analysis*, Fourth Edition. Unwin Hyman, London.

Moreno, R. Jr. and Bailey, D.G.F. (1989), Alternative Transport Fuels from Natural Gas, World Bank Technical Paper No. 96, Washington, D.C.

Mumme C. H. et al. (1992), The World Bank, Mexico: Transport Air Quality Management in the Mexico City Metropolitan Area Sector Study (Green Cover Report).

OECD (1991) The state of the environment. Annual report. Paris, Organization for Economic Cooperation and Development.

Ostro, B. (1994) *Estimating Health Effects of Air Pollution, a Methodology with an Application to Jakarta*. PROPE Research Project.

Ostro, B. (1983) The effects of air pollution on work loss and morbidity. *J. Env. Econ. and Man.*, 10:371-382.

Parikh, J.K. and J.P. Painuly (1991) *Compressed Natural Gas (CNG) in Transport Sector: A techno-economic assessment for India* (draft). Indira Gandhi Institute for Development Research, Bombay.

PT. Mojopahit Konsultama (1991) *Assessment of transportation growth in Asia and its effects on energy usage, environment and traffic congestion.* International Institute for Energy Conservation.

Shin, E, R. Gregory, M. Hufschmidt, Y-S Lee, J.E. Nickum, C. Umetsu (1992) *Economic valuation of urban environmental problems.* The World Bank, Washington D.C.

Tharby, R.D., W. Vandenhengel, and S. Panich (1992) *Transportation emissions and fuel quality specification for Thailand* (draft report Febr. 1992), Monenco consultants Ltd.

U.S. EPA (1982s) Air quality criteria for oxides of nitrogen: final report. Research Triangle Park, NC.

U.S. EPA (1982b) Air quality criteria for particulate matter and sulfur oxides. Research Triangle Park, NC.

U.S. EPA (1985) Compilation of air pollutant emission factors, 4th ed. Research Triangle Park, NC, EPA (Environmental Protection Agency; AP-42).

U.S. EPA (1986a) Air quality criteria for ozone and other photochemical oxidants. U. S. Environmental Protection Agency - reprint. Washington D.C.

U.S. EPA (1986b) Second addendum to air quality criteria for particulate matter and sulfur oxides: assessment of newly available health effects information. Triangle Park, NC.

U.S. EPA (1986c) Air quality criteria for lead. Draft final. Research Triangle Park, NC.

UNEP (1988) *Awareness and Preparedness for Emergencies at Local Level: The APELL Process.* Draft Handbook, July 1988.

UNEP/GEMS (1991) *Urban air pollution.* United Nations, Nairobi.

UNEP/WHO (1994) GEMS/AIR methodology reviews, vol.1: Quality assurance in urban air quality monitoring. Nairobi, UNEP (WHO/EOS 94.1) (UNEP/GEMS 94.A.2).

UNEP/WHO (1994) GEMS/AIR methodology reviews, vol.2: Primary standard calibration methods and network intercalibrations for air quality monitoring. Nairobi, UNEP (WHO/EOS 94.2) (UNEP/GEMS 94.A.3).

USAID and USEPA (1990) *Ranking Environmental Health Risks in Bangkok, Thailand.* Volume I. Washington.

USAID and USEPA (1990b) *Ranking Environmental Health Risks in Bangkok, Thailand.* Volume II. Technical Appendices. Washington.

Vatavuk, W.M. (1990) *Estimating costs of air pollution abatement.* Lewis Publishers, Chelsea, Michigan.

Walsh M.P., (1993), "Diesel Vehicle Pollution Control, Current Status and Global Trends", Washington D.C.

World Bank (1986) *Urban Transport.* A World Bank Policy Study, Washington D.C.

World Bank (1992) World development report 1992. Development and the environment. New York, Oxford University Press.

World Bank (1993) The World Bank Group and the Environment 1993. Development and the Environment, World Development Indicators. Washington DC USA.

World Bank (1994) The World Bank Group and the Environment 1994. Making Development Sustainable, Washington DC USA.

World Bank (1995) The World Bank Group and the Environment 1995. Mainstreaming the Environment, Washington DC USA.

WHO (1976) Manual on urban air quality management. Ed. by M.J. Suess and S.R. Craxford. Copenhagen, World Health Organization (WHO Regional Publ.; European Series 1).

World Health Organization (1983) *Guidelines on studies in environmental epidemiology.* Environmental Health Criteria No. 27, Geneva.

World Health Organization (1987) *Air Quality Guidelines for Europe.* Copenhagen: WHO. (WHO Regional Publications, European Series No. 23).

World Health Organization and United Nations Environment Programme 1992: *Urban Air Pollution in Megacities of the World,* Blackwell Publishers, Oxford.

World Health Organization (1992) *International Statistical Classification of Diseases and Related health problems.* Vol. 3. Alphabetical Index. Geneva.

Chapter 4
Mesoscale atmospheric transport and diffusion processes

Zafer Boybeyi[a] & Sethu Raman[b]
[a]Science Applications International Corporation,
1710 Goodridge Drive,
McLean, VA 22101 USA
[b]Department of Marine, Earth and Atmospheric Sciences,
North Carolina State University,
Raleigh, NC 27695-8208 USA

Abstract

Mesoscale atmospheric processes are important for the short and long range transport and diffusion of pollutants. Mesoscale motions are not readily quantized either in detail or statistical manner because they depend on terrain characteristics and the surface inhomogeneities which themselves are not easily quantized. Meteorological observations in this scale are also not readily available because of widely spaced monitoring stations. Consequently, the understanding of mesoscale atmospheric transport and diffusion processes are still incomplete. However, in recent years various field experiments and numerical studies have been conducted that have enhanced our understanding of mesoscale transport and diffusion processes. Numerical models with higher resolution and better representation of land surface processes have also contributed to our increased knowledge.

1. Introduction

Atmospheric dispersion is the irreversible process by which a cloud of pollutant expands, mixes, and is diluted by atmospheric motions acting on the cloud. Atmospheric dispersion may be separated into two components: transport and diffusion. The transport and diffusion are the processes due to mean wind and unresolved turbulent velocity fluctuations, respectively.

Since atmospheric motions occur over a wide range of scales from a few millimeters (e.g., turbulence) to thousands of kilometers (e.g., planetary waves), all of these flow scales can affect the transport and diffusion of atmospheric pollutants to some degree. Therefore, one typically desires to capture not only the development and evolution of small-scale flow features but also their interaction with and influence upon the larger-scale flow features. To date, however, the majority of atmospheric dispersion experiments have been concerned with local-scale dispersion within a few kilometers from a source. Over these distances, both the mean wind and turbulence are often assumed to be steady and horizontally homogeneous.

More recently, there has been growing concern about estimating the long-range dispersion over synoptic-scale distances. Synoptic-scale eddies such as extratropical cyclones are the dominant mechanism for dispersion on these scales. Mesoscale atmospheric dispersion can be defined as the transport and diffusion of pollutants (over travel distances between 2 and 2000 km) between the local-scale dispersion and the synoptic-scale dispersion. Over mesoscale distances, plumes and/or puffs are not only relatively localized with respect to their environment, but are also forced on scales larger than their own. A large number of mesoscale flow modes, including terrain-forced, synoptically-forced mean mesoscale circulations, and mesoscale fluctuations active on a wide range of temporal and spatial scales, constitute the primary mechanism for dispersion on this scale. Also, local-scale turbulent diffusion is important for mesoscale dispersion as are the synoptic-scale atmospheric flow scales and processes. Therefore, over mesoscale travel distances, the assumption of steady and horizontally homogeneous flow can no longer be valid, nor can the mesoscale flow be viewed in isolation (Pasquill, 1974).

2. Definition of mesoscale

The highly variable and complex phenomena of atmospheric flow can be classified according to the physical scales of apparently coherent structures that appear generally or intermittently within the flow (e.g., synoptic-scale, mesoscale, and local-scale). This scale classification might suggest that the atmosphere is somehow quantized and the scales are discrete. In reality, of course, the spectrum of atmospheric motions is smooth and continuous between the limits imposed by the mean free path of molecules and the circumference of the Earth.

For example, the spectra of wind speed near the ground for two different meteorological conditions are shown in Fig. 2.1. These spectra are referred to as energy spectra, because they have units of kinetic energy per unit mass.

The spectrum indicated by a solid line shows a large peak near a period of 4 days, a small peak near a period of 1 min, and a reduced energy region in between these two peaks. The reduced energy region indicates the existence of a mesoscale gap. The portion of the spectrum to the right of the small peak is called the inertial subrange (Kolmogorov, 1941). Theoretically, energy in the inertial subrange is neither created nor destroyed but transferred to higher and higher frequencies until it is dissipated by friction into heat.

On the contrary, the second spectrum indicated by a dashed line does not show the existence of the mesoscale gap. This implies that energy is produced in this range by mesoscale motions or that energy produced on a larger scale is broken down into this scale. Note that the peaks in the spectra occur at frequencies having the greatest contribution to the overall variance and correspond to production of turbulent kinetic energy.

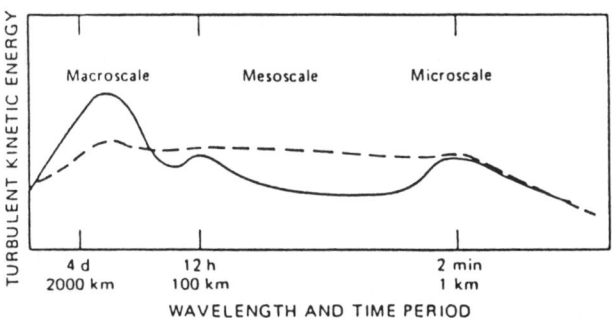

Fig. 2.1 Spectra of wind speed near the ground for two different synoptic conditions
(From D. Randerson, Atmospheric Science and Power Production,
USDOE Report TIC-27601, 1984)

The existence of such a mesoscale gap should be viewed as a near-surface feature resulting from boundary effects of the Earth's surface. The mesoscale gap has significant implications for (a) the predictability of large-scale atmospheric motions, (b) the parameterization of small-scale motions in terms of the characteristics of the large-scale motions, (c) the statistical stability of atmospheric turbulence, and hence (d) the predictability of mesoscale transport and diffusion processes.

For example, consider a cloud of pollutant released during periods with meteorological conditions corresponding to the two spectra presented in Fig. 2.1. Initially the two clouds would grow at the same rate by eddies, since the two time series have similar magnitudes for energy up to a scale of 1 km. As the clouds grow to mesoscale sizes, the cloud released during series 2 indicated by a dashed line would grow faster than the cloud released during series 1 indicated by a solid line, since more energy is available at these frequencies. As the clouds continue to grow to larger and larger sizes, their growth rates will be determined by the amount of energy at that scale. Differences in growth rates of the two clouds are, therefore, proportional to differences in the energy spectra.

As a somewhat arbitrary middle region of the atmospheric energy spectrum, the mesoscale can then be defined as that part of the spectrum between the synoptic-scale and local-scale. Further, the mesoscale can also be classified into three spatial decades as suggested by Orlanski (1975): the meso-γ scale (2-20 km), the meso-β scale (20-200 km), and the meso-α scale (200-2000 km). Although somewhat arbitrary, Orlanski's geometric definition has been widely adopted. Over these mesoscale distances, ubiquitous energy transfer processes and the special events occur locally and intermittently as a result of topographic forcing and mesoscale instabilities. Finally, the mesoscale atmospheric dispersion can be defined as the transport and diffusion of air pollutants over horizontal distances between 2 and 2000 km with time periods ranging from 1 to 48 hours.

3. Scale interaction

"Scale interaction" is one of the most popular phrases in the meteorological literature. In planetary-scale meteorology, scale interaction generally refers to the interactions between the zonal flow and a fairly limited set of waves that are quantized by the circumference of the Earth. In turbulence theory, it means the interactions among a continuous spectrum of eddies of all sizes. In mesoscale meteorology, scale interaction is the limited set of interactions among discrete scales such as the meso-γ, the meso-β, and the meso-α.

In a very simple system consisting of a slowly varying mean flow with a very weak perturbation upon it, the main scale interaction can be considered to be the influence of the mean flow upon the perturbation. As the perturbation becomes more substantial, it exerts an increasing influence on the mean flow. This generates other scales of motion as a result of secondary instabilities. Hence, the scale interactions become more and more numerous and the general degree of disorder becomes greater. In a highly nonlinear system such as fully turbulent flow, the scale interaction becomes very complex and numerous. Therefore, an explicit description of their interaction becomes problematic. One then may resort to stochastic treatments of the interactions. Perhaps the most difficult flows to deal with are ones for which the degree of disorder is too great to allow explicit treatment of the dynamics but too small to allow a simple stochastic treatment.

The question of the degree of disorder of mesoscale motions is a serious problem confronting mesoscale meteorologists.   Can the mesoscale be treated as finite number of waves interacting with each other and the mean flow?   Or can mesoscale be considered a continuous spectrum of eddies? According to Lilly (1983), the mesoscale is a region of disordered waves and eddies serving to shunt energy between the cumulus and cyclone scales. Mesoscale motions may also occur locally and intermittently to convert potential energy to kinetic energy on this scale.   If we assume that this is a qualitatively correct view, then we should ascertain the physical nature of the intermittent organized mesoscale motions and to determine the way in which energy, heat, and momentum are transferred by both the intermittent ordered motions and the more continuous disordered motions.

4. Basic features of mesoscale flow

In order to understand the transport and diffusion of atmospheric pollutants over mesoscale travel times and distances, it is first necessary to understand the behavior of atmospheric flows on these scales.   Until recently, the study of mesoscale meteorology lagged behind micrometeorology and synoptic meteorology due to a lack of suitable observational tools.   However, the advent in last few decades of new observing methods such as instrumented aircraft, constant-height balloons, Earth-orbiting satellites, dense networks of surface stations (mesonet), and remote-sensing electromagnetic probes (radar, sodar, lidar) has revolutionized mesoscale meteorology and provided an increasing data stream of mesoscale observations which improved our understanding of the basic features of mesoscale flow.

In the atmosphere, mesoscale energy is distributed over a variety of flow modes, including mean mesoscale circulations, mesoscale eddies, and internal gravity waves.   Flow dynamics are also quite variable and complex on this scale. Every atmospheric dynamical process operating on scales larger than molecular dissipation and smaller than the latitudinal variation of the Coriolis force can play a role in mesoscale circulations.   Mesoscale flows may be hydrostatic or nonhydrostatic.   Nonhydrostatic motions may contain significant features on scales ranging from several meters to several tens of kilometers with time scales of minutes to many hours.   Hydrostatic motions, in which the nonhydrostatic motions are embedded, have motion scales orders of magnitude larger than the nonhydrostatic motions.

Over mesoscale travel times and distances, density stratification in the atmosphere plays an important role in both flow dynamics and hence dispersion of pollutants. For example, a large number of processes (such as flow over orography, rising convective elements, shear instability, and frontal systems) can generate internal gravity waves due to the presence of density stratification. Internal gravity waves propagate mesoscale free oscillations. They can transport both momentum and energy over long distances. They are the simplest and most fundamental motions on the mesoscale, and are almost always present in the density-stratified free atmosphere. These waves can influence the atmospheric dispersion of pollutants in several ways. Their presence may result in mesoscale fluctuations in wind speed and direction depending upon the direction of propagation of the wave relative to the mean wind. Such periodic variations in wind direction may cause plume meander. These waves can also trigger flow instabilities such as Kelvin-Helmholtz instability and convective instability. These flow instabilities produce isolated patches of turbulence, even in the interior of stably stratified fluids. Thus, gravity waves can also influence the dispersion of pollutants by increasing turbulence intensity levels through these flow instabilities.

The presence of wind shear further complicates the flow dynamics and the dispersion of pollutants over mesoscale travel times and distances. Many flow instabilities occur partly as a result of the presence of wind shear, which may enhance turbulent diffusion by increasing turbulence intensities. For example, at night under stable conditions, the formation of a nocturnal surface inversion due to longwave radiational cooling decouples the nocturnal boundary layer from the remainder of the well-mixed daytime planetary boundary layer. In the nocturnal boundary layer, the wind speed decreases and the wind direction backs due to the new force balance between the pressure gradient force, the Coriolis force, and a weakened friction force. These changes in the wind direction and speed may result in pollutants at different levels being advected at different speeds (speed shear) or in different directions (directional shear). When morning arrives and the unstable daytime planetary boundary layer begins to grow, the shear-distorted pollutants mix in the vertical. Vertical turbulent mixing acts to reduce vertical wind shear and shear enhanced dispersion. Nevertheless, vertical shear will still be present mainly due to surface friction and baroclinicity, and its interaction with vertical diffusion results in simultaneous mixing.

Thermal and mechanical forcing due to mesoscale terrain inhomogeneities can generate externally forced mesoscale flow inhomogeneities such as upslope precipitation events, lee cyclogenesis, drylines, land and sea breezes, mountain-valley winds, and urban heat island circulations. Thermally generated mesoscale flows also include those due to horizontal contrasts in ground wetness and those due to spatial variations in vegetation/bare soil contrasts. Moist processes can generate mesoscale convective systems such as cumulus-scale convection, extratropical squall lines, mesoscale convective complexes, mesoscale cellular convection, tropical cyclones, mesoscale rainbands, tornadoes, and tornadogenesis. Synoptic-scale wave-wave interactions may produce higher wavenumber circulations or flow features. The atmospheric inertial mode can generate internally generated circulations such as fronts and jet streaks, buoyant instability, synoptic instability, Kelvin-Helmholtz instability waves, quasi-stationary convective events, and mesoscale structure of hurricanes. This wide range of mesoscale circulations and their associated mesoscale vertical ascents and descents can have a significant impact on the dispersion of pollutants. In addition, large-scale baroclinicity and large-scale vertical motions affect the local structure and the dynamics of planetary boundary layer. Latitude affects both the inertial period and diurnal forcing. Time of day affects the dynamics of the boundary layer, while time of year affects the diurnal forcing.

In summary, mesoscale atmospheric flows possess various mesoscale time and space scales resulting from the diurnal cycle, the atmospheric inertial mode, thermal and mechanical forcing due to mesoscale terrain inhomogeneities, and downscale energy transfer from the synoptic-scale to the mesoscale due to nonlinear flow interactions. Even for horizontally homogeneous flows over flat and uniform terrain, mesoscale frequencies such as the diurnal heating cycle and formation of a nocturnal low-level jet will usually be present. The transport and diffusion of atmospheric pollutants can be affected by this wide range of flow scales through variations in the mean transport wind, differential advection due to vertical and horizontal wind shear, and vertical mixing. Therefore, these mesoscale space-time flow scales should be represented accurately when the mesoscale dispersion processes are studied. Otherwise, significant components of mesoscale transport and diffusion may be neglected.

5. Basic features of mesoscale dispersion

In the case of local-scale dispersion, dilution occurs due largely to turbulent diffusion. Therefore, both the mean transport wind and turbulence at the local-scale may be considered steady and horizontally homogeneous. However, over mesoscale travel times and distances, this approximation of steady and horizontally homogeneous flow can no longer be valid. Instead, a large number of mesoscale flow modes active on a wide range of temporal and spatial scales constitute the primary mechanism for dispersion on this scale.

The relative importance of vertical diffusion vs. horizontal diffusion differs between mesoscale dispersion and local-scale dispersion. In the case of local-scale dispersion, vertical and horizontal diffusion will be comparable in magnitude under neutral or unstable conditions. On the contrary, in case of mesoscale dispersion, horizontal diffusion usually dominates vertical diffusion. This is due to the fact that once a tracer has mixed through the depth of the planetary boundary layer, additional dilution can only result from either further horizontal dispersion or from leakage out of the planetary boundary layer due to cloud venting, mesoscale venting, and large scale convergence. However, under stable conditions, vertical motions will be constrained by buoyancy forces, whereas horizontal motions will not be so constrained. Horizontal diffusion will thus dominate vertical diffusion in both local-scale dispersion and mesoscale atmospheric dispersion in stably stratified flows.

Another important feature of mesoscale atmospheric dispersion is the treatment of vertical motions. In local-scale dispersion, vertical motion can be neglected. Over mesoscale distances, however, the effects of mesoscale and larger scale vertical motions associated with convergence and divergence in the horizontal wind field may have a significant impact on dispersion. For example, mean vertical velocity values on the synoptic-scale may range from 1 cm/s downward in an intense anticyclone to a few cm/s upward in extratropical cyclones. Much larger vertical velocities can occur in thunderstorms, mesoscale convective systems, and along frontal surfaces.

Also, contrary to the idealized local-scale dispersion, non Gaussian cross plume and along wind surface concentration profiles are often observed at mesoscale travel distances. For example, in many mesoscale tracer experiments (Table 1), measured concentration fields showed multi surface concentration patterns downwind distances. Note that a simple Gaussian model for local-scale dispersion would predict a single ground level concentration maximum and then monotonically decreasing concentrations at greater downwind distances.

It can be concluded that mesoscale dispersion is an even more complex phenomenon than local-scale atmospheric dispersion due to the additional physical processes that may play a role and to the greater number of characteristic time and space scales. It would appear from this discussion that additional physical processes come into play in atmospheric dispersion at mesoscale space and time scales, even over flat, uniform terrain under simple synoptic conditions. Terrain forced mesoscale circulations over inhomogeneous terrain, can add further complexity to pollutant dispersion as can changing synoptic conditions. A comprehensive discussion on the basic features of mesoscale atmospheric dispersion can be found in Moran (1992).

6. Mesoscale flow scales

The atmosphere is characterized by phenomena whose space and time scales cover a very wide range. The space scales of these features are determined by their typical size or wavelength, and the time scales by their typical lifetime or period. Over mesoscale travel times and distances, a large number of mesoscale flow modes are possible with a wide range of attendant temporal and spatial scales. The implication, of course, is that mesoscale flow scales and flow structures must be considered and characterized, when mesoscale atmospheric dispersion is studied.

6.1 Mesoscale time scales

The influence of the Earth's surface is effectively limited to the lowest 10 km of the atmosphere in a layer called the troposphere. Over time periods of about one day this influence is restricted to a very much shallower zone known as the planetary boundary layer. The planetary boundary layer contains typically about ten per cent of the overlying mass of air. This layer is particularly characterized by well developed turbulence (or mixing) generated by frictional drag as the atmosphere moves across the rough and rigid surface of the Earth, and by the density stratification of the air which results from differences in temperature of the surface and the air. These differences in temperature arise over land primarily in the course of the daily cycle of radiative heating and cooling of the ground, but they may also arise from overflowing of air from warmer or cooler regions of the Earth. The boundary layer receives much of its heat and all of its water through this turbulence process. Undoubtedly, effective dispersion of atmospheric pollutants released into the atmosphere near the ground depends largely on mixing processes on a variety of scales. This mixing is a direct consequence of turbulent and convective motions generated in the planetary boundary layer.

In the aerodynamic sense, the boundary layer is simply the layer from which momentum is extracted and transferred downward to overcome surface friction. This layer has important variations in depth as well as in mixing efficiency, arising from the control on vertical transport by the density stratification. The height of the boundary layer is not constant with time, it depends upon the strength of the surface generated mixing. By day, when

Earth's surface is heated by the Sun, there is an upward transfer of heat into the cooler atmosphere. This vigorous thermal mixing enables the boundary layer depth to extend to about 1 to 2 km. Above the planetary boundary layer, a stably stratified free atmosphere usually forms in which turbulent motion is suppressed. Conversely by night, when the Earth's surface cools more rapidly than the atmosphere, there is a downward transfer of heat. This tends to suppress mixing and the boundary layer depth is confined to a shallower layer than in daytime.

Within the planetary boundary layer, these changes in vertical stability and vertical mixing resulting from the diurnal heating cycle modify the wind speed, wind direction, and turbulence profiles even over flat and uniform terrain. One striking example of this is the frequent formation of a low-level jet over land on clear nights. Although the timing of the formation and subsequent erosion of nocturnal jet is controlled by the diurnal cycle, the jet itself can also be explained as an inertial oscillation. Thus, two natural mesoscale time scales need to be considered for travel times of more than one hour; the 24-hour diurnal period and the latitude dependent inertial period $2\pi/f$, where f is Coriolis parameter. The inertial period may be longer than, equal to, or shorter than the diurnal period depending on the latitude.

For thermally-forced mesoscale circulations, the natural circulation time scale is the forcing time scale, that is, the duration of heating or cooling. This time scale may vary from 1 to 24 hours depending upon latitude and season. Short-duration forcing is less likely to generate a mesoscale atmospheric response than longer duration forcing. However, some shallow circulations such as drainage winds can form in as little as one or two hours. In contrast, mesoscale circulations resulting from the mechanical forcing of stably-stratified flows by topographic features will have the time scale of the larger scale flow even though they are mesoscale in spatial extent. Still other mesoscale time scales will be associated with internally-forced mesoscale phenomena such as fronts, jet streaks and latent heat driven phenomena (e.g., mesoscale convective complexes or mesoscale rainbands). These time scales will often be comparable to or larger than the inertial or diurnal periods, but they can also be shorter, especially if internal gravity waves are important.

6.2 Mesoscale space scales

Atmospheric flows are seldom steady over mesoscale time periods. Therefore, they are not usually horizontally homogeneous over mesoscale distances. Many mesoscale circulations are atmospheric responses to externally-imposed boundary forcing such as mountains, plateaus, and valleys. The presence of such topographic features can result in mesoscale flow phenomena such as mountain waves, lee waves, vortex shedding, upstream blocking, channeled flow, slope winds, and mountain-valley circulations. Variations in other terrain properties may also affect atmospheric flows. Horizontal inhomogeneities in roughness length or surface temperature can generate internal boundary layers, and mesoscale atmospheric circulations such as land-sea breezes and urban heat island circulations (e.g., Boybeyi and Raman, 1992).

These various surface-forced mesoscale circulations possess a number of natural spatial scales. In case of mechanically-forced circulations, the horizontal and vertical scales will depend on obstacle height h, obstacle width l, horizontal wind speed V, vertical stratification represented by the Brunt-Vaisala frequency N, and density scale height $H = \rho_o (\partial \rho_0 / \partial z)^{-1}$, where $\rho_0(z)$ is the base state density. In general, the vertical scale of mechanically-forced mesoscale circulations ranges between the obstacle height and the density scale height, while the horizontal scale lies between the horizontal scale of the obstacle and the Rossby radius of deformation based on obstacle height Nh/f. In the case of thermal forcing, thermally-forced mesoscale circulations generally tend to be shallow so that the planetary boundary layer height z_i is a relevant vertical scale within the Rossby radius of deformation Nz_i/f as the limiting scale. However, differences in surface heating due to surface inhomogeneities may also trigger moist convection and mesoscale convective clouds. In this case, convection tends to be cumulus scale and may develop throughout the depth of the troposphere.

Spatial variations in atmospheric fields may also arise due to internal flow dynamics and instabilities. Mesoscale circulations can be generated by synoptic-scale dynamical or thermal forcing. Examples of such circulation systems include frontal and dryline circulations, density currents, squall lines, jet streaks, and mesoscale gravity waves. In addition to geostrophic adjustment (and flow over topography), internal gravity waves with horizontal wavelengths ranging from several hundred meters to several hundred kilometers can be generated by Kelvin-Helmholtz instability, frontal surges, convective activity, and restratification processes. Quasi-horizontal vertical modes may be generated by three-dimensional turbulence decay and upscale energy transfer in a stratified fluid. Synoptic-scale baroclinicity can also modify vertical profiles of mean wind and turbulence within the planetary boundary layer by imposing additional vertical shear. Mesoscale or synoptic-scale vertical motions resulting from internal flow dynamics may concentrate pollutants near the Earth's surface or lift them out of the planetary boundary layer. Clearly, mesoscale space scales can be quite complicated, and can directly influence mesoscale dispersion.

7. Dispersion scales

Besides the mesoscale flow scales just discussed above, the transport and diffusion of a passive scalar introduce additional spatial and temporal scales.

7.1 Source configuration

A number of important characteristics of the emissions have a significant influence on the resulting air pollution. The rate of emission and the physical and chemical nature of the pollutants are essential to the determination of the amount and type of pollutant. It is also important to know certain other characteristics of the source including the shape of the pollutant source, the duration of the releases, and the effective height at which the injection of pollutants occurs.

A pollutant source may be categorized according to its geometry as a point source, line source, area source, and volume source. For example, the most important point source is the chimney stack which can be the originator of very concentrated and harmful materials. A busy highway is the most common example of a line source where it is assumed that the integration of the exhaust emissions from many separate vehicles constitutes a continuous output along its length. Cities are the paramount area sources wherein it is convenient to lump together the individual emissions from a multiplicity of small sources (e.g., houses, etc.) to give an areal average. It should be pointed out that these source designations depend upon the scale of the study. For example, on a synoptic-scale, cities might appear as point sources. Therefore, pollutant sources are often approximated as point sources. This approximation becomes increasingly valid as the distance to the location at which pollutant concentration is desired becomes large with respect to the actual source dimensions. Hence, the size and the shape of the source introduce new spatial scales. The location of the source may do the same. For example, the distance from a pollutant source to a major terrain inhomogeneity such as a large body of water or a mountain range may be important. In the case of an elevated source, the heights of the source above the ground and below the planetary boundary layer capping inversion are also important spatial scales.

On the other hand, release time or the duration of release constitutes an important dispersion time scale. For example, instantaneous releases (puffs) and continuous releases (plumes) must often be treated quite different. This release time scale can be important because the relative diffusion of a cloud is scale dependent. The length of the release time relative to the travel time to a downstream receptor also affects the fraction of time that non-zero concentration levels measured at that receptor are quasi-steady.   If the release time is large compared to the travel time, then unsteady concentration levels associated with the passage of the leading and trailing edges of the pollutant cloud constitute only a small fraction of the total concentration trace.   If, however, the release is terminated well before the leading edge of the pollutant cloud reaches the downstream receptor site, than the concentration trace will likely ramp up to a maximum value and then ramp down with no period of quasi-steady concentration levels (e.g., Draxler, 1988).

7.2 Instantaneous releases

The instantaneous release or puff release is the conventional approximation to the type of release associated with a very short venting of material to the atmosphere.   The puff, once formed, moves away from the source with a speed and direction determined by the wind at the moment of release.   The mean speed and direction of the puff can be expected to change from the original values during its travel as the wind pattern in which it is embedded changes with time.   As the puff moves, it will expand about its center owing to the action of turbulent fluctuations.

For puff releases, the natural frame of reference is a coordinate system that moves with the puff center of mass, i.e., a Lagrangian reference frame, and the puff characteristic of greatest interest is the size of the puff.   The simplest representation of the diffusion of such a puff is the rate of separation of a pair of particles released simultaneously from the same source.   For this reason, the growth of a puff about its center of mass is often referred to as two-particle or relative diffusion.

The manner in which a puff release disperses is essentially determined by the size of the puff relative to the spatial scale of the dominant motions of the flow field in which it is embedded.   Idealized transport and diffusion patterns are shown schematically in Fig. 7.1.   The left side of Fig. 7.1a shows a typical puff in an atmosphere containing many small atmospheric motions called eddies.   Since the size of the eddies is much smaller than the size of the puff, these eddies tend to slowly diffuse the puff and make it grow larger in time.   On the contrary, when the puff is much smaller than the eddies, as shown in Fig. 7.1b, the motions tend to transport the cloud rather than diffuse it.   When the cloud and eddies are nearly the same size, the cloud grows rapidly.   Thus, a fundamental property of relative diffusion is that the rate of growth of a puff depends directly on the size of the puff.   And a puff released into the atmosphere will always grow larger since there almost always will be eddies smaller than or of the same size as the puff.

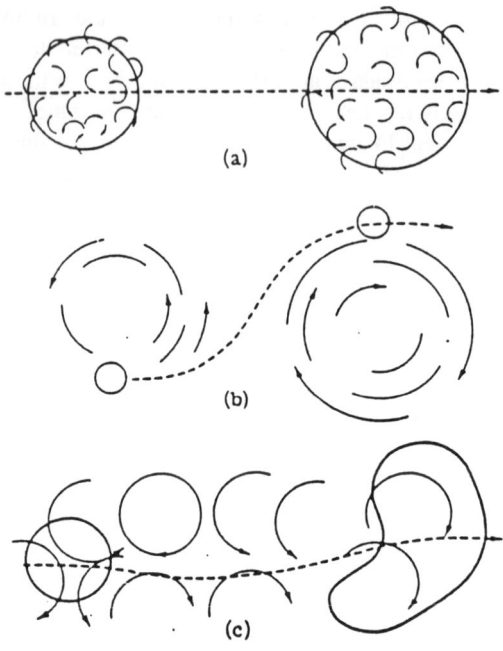

Fig. 7.1 Idealized dispersion patterns; (a) a large puff in a uniform field of small eddies, (b) a small puff in a uniform field of large eddies, (c) a puff in a field of eddies of the same size as the puff (From D. H. Slade, *Meteorology and Atomic Energy*, USAEC Report TID-24190, 1968).

7.3 Continuous releases

A continuous plume may be considered to be made of an infinite number of puffs released from a fixed or stationary source sequentially with a vanishingly small time interval between puffs. The resulting downstream plume remains connected to the source. Thus, the natural reference frame for a plume release is a coordinate system that is fixed with respect to the source. In such an Eulerian reference frame, it is easy to vary the period of time over which the resulting pollutant plume is sampled. Initially each puff moves with the wind direction at the moment of release. Linear dimensions of the plume perpendicular to the plume axis are often given in terms of the standard deviation of the concentration distribution since the average cross-plume distribution is usually close to a normal curve with boundaries at infinity.

The left side of Fig. 7.2 denotes, schematically, the boundaries of a plume as it would appear at any given instant or as it would appear having been averaged over 10 min or 2 hours. The right side of the figure represents the distributions of the concentration for each of the three plume types. The snapshot shows an irregularly varying plume, which gradually expands with distance from the source. The corresponding concentration distribution is irregular and quite peaked. The 10 min average plume shows a smooth and regular boundary that can be considered as the envelope containing all the short-period fluctuations of the plume. The average concentration distribution perpendicular to the axis of such a plume has been found to resemble a Gaussian distribution about the center line, as shown on the right side of the figure. The 10 min cross-plume average concentration distribution is broader and smoother than in the instantaneous case but has a lower peak value. The 2 hours plume is wider with an even flatter average concentration pattern.

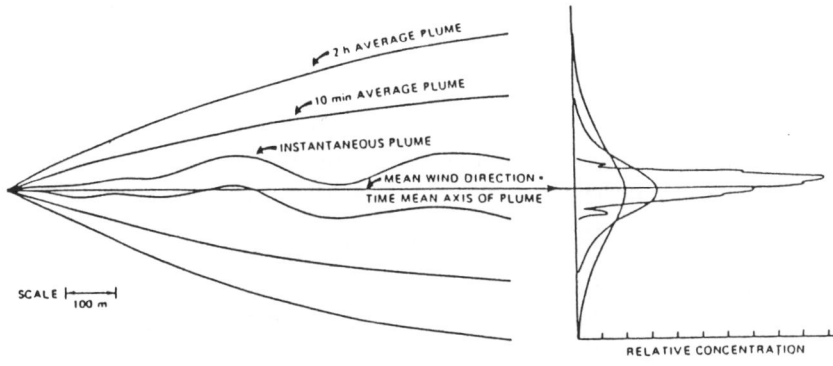

Fig. 7.2 The diagram on the left represents the appropriate outlines of a plume observed instantaneously and of plumes averaged over 10 min and 2 hr. The diagram on the right shows the corresponding cross-plume distribution patterns (From D. H. Slade, *Meteorology and Atomic Energy*, USAEC Report TID-24190, 1968).

If the time-averaged diagrams of the plume were extended to distances quite far from the source, the boundaries of the time-smoothed plume would meander because the longer length of plume would be under the influence of eddies that are quite large in area. The averaging time used originally therefore would be too short to show a time-averaged picture of these larger fluctuations. A longer time average appropriate to this greater distance would, again, be too short for distances greater yet. It is important to recognize that fluctuations in the wind that are larger than the plume dimensions tend to transport the plume intact (a process known as meander) whereas those that are smaller tend to tear it apart resulting in further mixing, growth, and dilution. Thus for a plume, all scales of motion contribute initially to mixing, but as the plume reaches greater distances and grows in size, larger turbulent eddy sizes become effective in diffusing the plume, and smaller eddies become increasingly ineffective. For example, on a scale of days, synoptic-scale flow fluctuations can change the direction of plume transport.

Observing or sampling time is thus an important quantity when studying continuous releases. This dependence of plume concentration on sampling time is the result of averaging over flow fluctuations or eddies with time scales shorter than the sampling period. Since eddy time scales are normally directly proportional to spatial scale, increasing the sampling time incorporates a greater and greater proportion of all eddy contributions and hence a greater range of plume displacements.

8. Fundamentals of atmospheric turbulence

It is obvious from the above discussions that a knowledge of fundamentals of atmospheric turbulence and averaging techniques is essential to understanding of mesoscale transport and diffusion processes. Turbulence refers to the chaotic nature of many flows, which is manifested in the form of irregular, random fluctuations in velocity, temperature, and scalar concentrations around their mean values in time and space. There is a very easy way to isolate the large-scale variations from the turbulent ones. By averaging the wind speed measurements over a time period (say 30 minutes to one hour), we can eliminate or average out the positive and negative deviations of the turbulent velocities about the mean. The classical Reynolds treatment of three-dimensional turbulence is based on the concept of a rapidly fluctuating turbulent component superimposed on a stationary, or at least slowly varying mean flow as follows:

$$\xi = \bar{\xi} + \xi' , \tag{1}$$

where ξ could represent such variables as velocity, temperature, or scalar concentrations, the overbar denotes an average of some sort over some scale, and the prime denotes a fluctuation from that average.

8.1 Averaging techniques

By introducing averaging with Eq. (1), we can concentrate on mean values and ignore many flow details. However, an averaging operator must be chosen, bearing in mind that the use of different operators may lead to different decompositions of the flow and hence to different mean and turbulence fields.

The main types of averages are time averages, space averages, space-time averages, and ensemble averages. The various types of averages are appropriate to different situations. Time averages are commonly used for measurements made over a sampling interval at a fixed point in space by in situ instruments such as hot-wire anemometers or instrumented towers. Space averages are more demanding for practical applications, since they may require knowledge of instantaneous variable values over three spatial dimensions. Nevertheless, some measurements made by remote-sensing techniques such as photography, radar, sodar, and lidar yield space averages. Accordingly, measurements from remote-sensing instruments over a finite sampling time will yield space-time averages. On the other hand, a true ensemble average taken over an infinite number of realizations only has theoretical utility, although, an ensemble average can also be calculated based on only a finite number of realizations. Since, the modern theory of turbulence has adopted a statistical theory approach, dynamical variables in a turbulent flow are assumed to be random fields and ensemble averages are used instead of time or space averages.

8.2 Basic features of atmospheric turbulence

The usual definition of kinetic energy is $KE = 0.5 \, m \, V^2$, where m is mass. When dealing with a fluid such as air, it is more convenient to talk about kinetic energy per mass unit. It is also convenient to partition the kinetic energy of the flow into a portion associated with the mean wind (MKE), and a portion associated with the turbulence (TKE). By taking advantage of the mean and turbulent parts of velocity introduced above, we can immediately write the desired equations:

$$\frac{MKE}{m} = \frac{1}{2}\left(\overline{u}^2 + \overline{v}^2 + \overline{w}^2\right), \tag{2}$$

$$e = \frac{1}{2}\left(u'^2 + v'^2 + w'^2\right), \tag{3}$$

where e represents an instantaneous turbulence kinetic energy per unit mass. Rapid variations of e with time can be expected as we measure faster and slower gusts. By averaging over these instantaneous values, we can define a mean turbulence kinetic energy that is more representative of the overall flow:

$$\frac{TKE}{m} = \frac{1}{2}\left(\overline{u'^2} + \overline{v'^2} + \overline{w'^2}\right) = \overline{e} \ . \tag{4}$$

We can immediately recognize that mean turbulence kinetic energy is nothing more than the summed velocity variances divided by two. Therefore, using prognostic equations for the sum of velocity variances and dividing by two easily gives us the TKE budget equation written in summation notation (see Stull 1988 for more information about the derivation of the TKE budget equation):

$$\frac{\partial \overline{e}}{\partial t} + \overline{U}_j \frac{\partial \overline{e}}{\partial x_j} = +\delta_{i3}\frac{g}{\overline{\theta}_v}\left(\overline{u_i'\theta_v'}\right) - \overline{u_i'u_j'}\frac{\partial \overline{U}_i}{\partial x_j} - \frac{\partial\left(\overline{u_j'e}\right)}{\partial x_j} - \frac{1}{\rho}\frac{\partial\left(\overline{u_i'p'}\right)}{\partial x_i} - \varepsilon \ , \tag{5}$$

where first term represents local storage or tendency of TKE, second term the advection of TKE by the mean wind, third term the buoyant production or consumption, fourth term the mechanical or shear production/loss term, fifth term the turbulent transport of TKE which describes how TKE is moved around by the turbulent eddies, sixth term the pressure correlation term that describes how TKE is redistributed by pressure perturbation, and the last term the viscous dissipation of TKE.

The above equation shows that there are two production mechanisms for turbulent kinetic energy in atmosphere. In an initially nonturbulent flow, the onset of turbulence may occur suddenly through an instability mechanism acting upon the naturally occurring disturbances in the flow. Since turbulence is transported downstream in the manner of any other fluid property, repeated instabilities are required to maintain a continuous supply of turbulence. Once turbulence is generated and becomes fully developed, the instability mechanism is no longer required. This is particularly true in the case of shear flows, where shear provides an efficient mechanism of converting mean flow energy to turbulent kinetic energy. Similarly, in unstably stratified flows, buoyancy provides a mechanism of converting potential energy of stratification into turbulent kinetic energy. Thus, in order to maintain turbulence in the atmosphere, one or more the production terms must be active continuously. In the daytime planetary boundary layer, turbulence is produced both by shear and buoyancy, while shear is the only effective way of producing turbulence in the nocturnal boundary layer.

There is also a continuous dissipation of turbulent kinetic energy, and there may be transport of energy from or to other regions of flow. Physically, this means that turbulence will tend to decrease and disappear with time, unless it can be generated locally or transported by mean, turbulent, or pressure perturbation processes. Also note that in stationary or steady-state conditions, the rate at which the energy is dissipated is exactly equal to the rate at which the energy is supplied from mean flow into turbulence.

In large Reynolds number flows, the main part of the mechanical and thermal production of turbulent kinetic energy takes place at large eddies, while almost all of it is eventually dissipated by small eddies. The largest boundary layer eddies scale to the depth of the boundary layer, that is, 100 to 3000 m in diameter. These are the most intense and energy containing eddies, while smaller eddies, on the order of a few millimeters in size, are very weak because of the dissipating effects of molecular viscosity. The transfer of energy from large energy containing eddies to small energy dissipating eddies occurs through a cascade type process involving the whole range of intermediate size eddies.

The physical picture behind this argument is that buoyancy and mean shear produce eddies of large geometrical size. These eddies are dynamically unstable and break into smaller ones, which break into even smaller ones, etc., until the eddies become so small that they are dominated by viscous forces and dissipated rapidly. This may be considered to be a qualitative picture of turbulence structure and the concept of energy transfer down the scale. Hence, TKE is not a conserved quantity.

The turbulence kinetic energy is one of the most important quantities used to study the planetary boundary layer. The motions in the planetary boundary layer are almost always turbulent, while it may be intermittent and patchy in the upper part of the planetary boundary layer and is sometimes mixed with internal gravity waves. Turbulence is responsible for efficient mixing and exchange of mass, heat, and momentum between the atmosphere and the Earth's surface throughout the boundary layer. For example, the radiation balance and the heat energy balance near the Earth's surface are significantly effected by the sensible and latent heat exchange processes between the Earth's surface and the atmosphere. Turbulent transfer of momentum between the Earth and atmosphere acts as a sink of atmospheric momentum in order to exert frictional resistance to atmospheric motions. The atmosphere receives virtually all of its water vapor through turbulent exchanges near the surface. Besides water vapor, there are other important exchanges of mass within the boundary layer involving a variety of gases and particulates. Through efficient diffusion of the various pollutants released near the ground and mixing them throughout the planetary boundary layer and parts of the lower troposphere, atmospheric turbulence prevents the fatal poisoning of life on Earth. Therefore, turbulent exchange processes within the planetary boundary layer have profound effects on the local weather and the mesoscale transport and diffusion of pollutants released near the surface.

9. Observational studies of mesoscale atmospheric dispersion

Observations of mesoscale atmospheric dispersion provided the impetus for many theoretical and numerical studies. Table 1 lists some of the major mesoscale tracer experiments.

Relatively few mesoscale atmospheric dispersion observations exist compared to the large number of observations of short-range dispersion experiments made. There are a number of reasons for this difference, including an early lack of interest on the part of both the scientific community and various funding agencies, a lack of suitable technologies for observing atmospheric dispersion over larger scales, and the high costs and organizational and logistical demands associated with mounting a mesoscale atmospheric dispersion experiment. The majority of the experiments listed in Table 1 have been carried out in the last decade, reflecting both the recent development of suitable tracers and analysis techniques and the heightened interest of various research groups and funding agencies in mesoscale atmospheric dispersion.

A well designed mesoscale atmospheric dispersion field program requires coordinated and extensive sensor networks deployed over a large area to measure both meteorological variables and tracer concentrations. A variety of tracer gases have been used in mesoscale tracer experiments, including sulfur hexafluoride (SF6), halocarbons (CBrF3, CBr2F2), radioactive gases (Ar-41, Kr-85), deuterated methanes (Me-20, Me-21), and perfluorocarbons (PMCH, PDCH, PTCH, PDCB, PDCH, PMCP).

An ideal tracer should be non-toxic, non-reactive, non-depositing, inexpensive, and detectable at very low concentration. The perfluorocarbons are the closest to being ideal tracers. Especially perfluoromonocyclohexane (PMCH) is the most popular tracer currently being used for mesoscale tracer experiments. It is inert, harmless, and can be accurately measured at its ambient background concentration. PMCH is dynamically passive in the sense that it does not carry meteorologically significant energy and does not affect the atmospheric flow field. It does not react chemically in the gas phase, has a negligible deposition velocity at the Earth's surface, and is not appreciably soluble in water so that it is not removed from the gas phase by clouds, haze, or precipitation scavenging. However, all perfluorocarbons are expensive to produce and require elaborate analytical equipment to make concentration measurements.

Table 1: The list of some of the mesoscale tracer experiments

Experiment Name	Location	Period mo-yr	Tracer	Domain Size	Release Duration
Great Plains	Central US	7-80	SF6, PMCH, PDCH	600 km	3 h
SEADEX	NE Wisconsin	5-82 6-82	SF6	30 km	5 h
ACURATE	U.S. Atlantic Atlantic	3-82 9-83	Kr-85	1100 km	12 h
CAPTEX	NE U.S. SE Canada	9-83 10-83	PMCH	1100 km	3-6 h
METREX	Washington, DC	11-83 11-84	PMCP, PMCH, PDCH	90 km	8 h-1 mo
DOPPTEX	San luis Obispo, CA	8-86 9-86	SF6, CBrF3	50 km	3-8 h
WHITEX	NE Arizona SE Utah	1-87 2-87	Me-20	250 km	2 mo
ETEX	Europe	10-94 11-94	SF6	2000 km	12 h

10. Theoretical studies of mesoscale atmospheric dispersion

The problem of turbulent diffusion in the atmosphere has not yet been uniquely formulated in the sense that a single basic physical model capable of explaining all the significant aspects of the problem has not yet been proposed.   However, there have been a number of attempts to apply concepts developed in the theory of small-scale dispersion to the mesoscale case. There are available two alternative approaches; the gradient transport theory and the statistical theory. Diffusion at a fixed point in the atmosphere, according to the local gradient transport theory, is proportional to the local concentration gradient. Consequently, it could be said that this theory is Eulerian in nature in that it considers properties of the fluid motion relative to a spatially fixed coordinate system. On the other hand, statistical diffusion theory considers motion following fluid particles and thus can be described as Lagrangian. This section reviews briefly these theories relevant to the mesoscale case.

10.1 Gradient transport theory

No general approach to the solution of problems in turbulence exists. The equations of motions have been analyzed in great detail, but it is still impossible to make accurate quantitative predictions without relying heavily on empirical data. Statistical studies of the equations of motion always lead to a situation in which there are more unknowns than equations. This is the closure problem of turbulence theory: one has to make assumptions to make the number of equations equal to the number of unknowns.

In order to close the set of equations, the variances and covariances must be either specified in terms of other variables or additional equations must be developed for the same. In the latter approach, the closure problem is only shifted to a higher level in the hierarchy of equations that can be developed. However, a more widely used approach has been based on the assumed analogy between molecular and turbulent transfers. It is called gradient-transport approach, because turbulent transport or fluxes are sought to be related to the appropriate gradients of mean variables (velocity, temperature, concentration, etc.). For example, in order to close the conservation of mass of a single pollutant species C, the covariance term can be approximated as follows:

$$\overline{w'C'} = -K\frac{\partial C}{\partial z}, \tag{6}$$

where K is the turbulent eddy diffusivity. Then, the closure problem is reduced to specify the K values. It should be, however, recognized that the above gradient transport relation is not the expression of any sound physical law, but only an intuitive assumption of similarity or analogy between molecular and turbulent transfer. Eddy diffusivities are usually several orders of magnitude larger than their molecular counterparts, indicating the dominance of turbulent mixing over molecular exchanges. More importantly, eddy diffusivities cannot simply be regarded as fluid properties; these are actually turbulence or flow properties, which can vary widely from one flow to another and from one region to another in the same flow.   Eddy diffusivities show no apparent dependence on molecular properties, such as mass density, temperature, etc., and have nothing common with molecular diffusivity, except for the dimensions.   In spite of these limitations of the implied analogy between molecular and turbulent diffusion, the gradient transport theory is quite useful and is widely used in practice.

10.2 Statistical theory

There are two basic forms of statistical theories of turbulent diffusion: (1) diffusion from a fixed emitting source and (2) diffusion of a puff about its center. The former, which is often referred to as absolute diffusion or single-particle diffusion, requires an Eulerian analysis framework while the latter, which is often referred to as relative diffusion or two-particle diffusion, requires a Lagrangian framework.   What actually distinguishes these two basic types of turbulent diffusion from each other is the analysis reference frame rather than the release type. This distinction should become clear from the following discussion.

10.2.1 Absolute diffusion

The basis of the statistical theories of diffusion is the study of particles of fluid as they travel in the turbulent flow. The pioneering study in this area was done by Taylor (1921) in an analysis that provides the foundation for much of our current understanding of turbulent diffusion including theoretical development and guidance for practical models. Taylor restricted his analysis to homogeneous, stationary turbulence and analyzed the continuous motion of a particle introduced into this flow. In one dimension a very fundamental statement about the particle is that its distance from the origin is due to its accumulated motion since release. This makes it possible to subtract the mean velocity and consider only the turbulent departures from the mean. This leads to the expression for the distance from the origin.

$$y(t) = \int_0^t v'(t')dt', \tag{7}$$

where v is the fluctuating turbulent velocity. Using the above equation, we can derive useful information by repeating the release and tracking process a large number of times and by preserving the second moment of the particle distribution and by squaring both sides of the above equation and then averaging over the ensemble of repeated experiments. This leads to

$$\overline{y^2}(t) = 2\overline{v'^2} \int_0^t \int_0^{t'} R_L(\xi)d\xi dt',$$

(8)

where the overbar denotes the ensemble average and the function $R_L(\xi)$ is the Lagrangian autocorrelation coefficient

$$R_L(\xi) = \frac{\overline{v'(t)v'(t+\xi)}}{\overline{v'^2}}.$$

(9)

Theories and measurements of turbulence have provided useful information on the form of $R_L(\xi)$. In particular, it must be unity at $\xi = 0$ and must approach zero at very large ξ. These two properties yield useful information from Eq. (8) for the limits of the second moment for small and large travel times. At small times, $R_L(\xi)$ is approximately constant at unity, and Eq. (8) can be simply integrated to give

$$\overline{y^2}(t) \approx \overline{v'^2}t^2.$$

(10)

At long times, since $R_L(\xi)$ converges to zero, the inner integral in Eq. (8) reaches a fixed value, K, and

$$\overline{y^2}(t) \approx 2Kt,$$

(11)

where K is the eddy diffusivity coefficient. Eq. (8) may be converted to a form that yields insight into the role of eddy sizes by using the Fourier transform relation between $R_L(\xi)$ and the Lagrangian energy spectrum

$$\overline{y^2}(t) = \overline{v'^2}t^2 \int_0^\infty F(n)\frac{\sin^2(\pi nt)}{(\pi nt)^2} dn.$$

(12)

The above equation expresses the particle position variance in terms of a filtered spectrum. All scales of motion contribute to the spread at all times. However, the second factor in the integral determines a weighting function that emphasizes certain ranges of eddy size. At small t, a wide range of frequencies, n, contribute to the integral, whereas at large t, only small frequencies (large eddies) have significant weighting in the calculation. We can interpret the last result in the following way. Close to a continuous source, all scales of turbulent eddies contribute to plume spread. Since the plume is narrow, small eddies serve to deform and tear it apart, while large eddies account for a meander of the entire plume, thus producing an effectively wider time-averaged plume. At long downwind distance the small eddies merely move material around within a wide plume while the large eddies still provide significant local deformation.

10.2.2 Relative diffusion

Although the absolute diffusion theory proposed by Taylor is probably the single most influential theoretical development in turbulent diffusion, it was recognized quite early that many realistic problems require information on diffusion relative to the centroid of a puff or plume. Brier (1950) and Batchelor (1952) investigated the relative diffusion problem in terms of the simplest ensemble a pair of particles. Taylor's method applied to the velocity difference between the two particles leads to expression

$$\overline{Y_i(t)Y_j(t)} = \overline{Y_i(t_0)Y_j(t_0)} + \int_{t_0}^{t}\int_{t_0}^{t}\overline{w_i(t')w_j(t'')}dt'dt'' \,, \tag{13}$$

where the capital letter notation refers to particle separations in the i and j coordinate directions and the w terms are velocity differences between the two particles. For analysis purposes, we can select only one component (e.g., $i = j = 1$) and expand the velocity difference terms in the integral and Eq. (13) becomes

$$\overline{Y^2(t)} = \overline{Y^2(t_0)} + 2v'^2\int_{t_0}^{t}\int_{t_0}^{t}R_L(t''-t')dt'dt'' - 2\int_{t_0}^{t}\int_{t_0}^{t}\overline{v_1'(t')v_2'(t'')}dt'dt''. \tag{14}$$

The subscripts on the velocities here refer to particles 1 and 2. The major differences between Eqs. (14) and (8) are the dependence on the initial separation between the two particles and the final integral term that depends on a correlation between particle 1 at one time and the particle 2 at another time. The latter term has been a major block to the progress of the Taylor approach to relative diffusion because it is difficult to evaluate for several reasons. It is nonstationary since, in general, the relative velocities change as the separation changes. Hence the relative diffusion rate depends on an additional time scale.

We can draw some conclusions from Eq. (14) regarding the effective scale of eddies for growth of an ensemble of particle pairs relative to a local centroid. Eddies much larger than $\overline{Y^2(t)}$ will displace pairs of fluid points together, whereas eddies smaller than $\overline{Y^2(t)}$ will serve to separate particles.

Hence the small eddies have greater efficiency. On the other hand, the role of the small eddies is diminished in a manner similar to fixed point diffusion, and we are left with the somewhat intuitive result that eddies on the size scale of the particle ensemble (or puff) are the most effective at spreading the puff.

11. Numerical studies of mesoscale atmospheric dispersion

Mesoscale dispersion models are either Eulerian or Lagrangian in their basic formulation. The Eulerian models adopt a fixed frame of reference, usually the surface of the Earth, while the Lagrangian models employ a moving frame of reference, usually that of an air parcel of interest.   In general, Eulerian models contain computational grid cells, while Lagrangian models are grid-free. A comprehensive review of dispersion modeling methods is given in Zannetti (1990).

11.1 Eulerian dispersion models

The Eulerian models are based on the conservation of mass of a single pollutant species

$$\frac{\partial C}{\partial t} = -V.\nabla C + D\nabla^2 C + Q , \tag{15}$$

where $C(x,y,z,t)$ is the pollutant concentration, D is the molecular diffusivity, Q is the source and/or sink term, ∇ is the gradient operator, and ∇^2 is the Laplacian operator.   It can be assumed that the velocity V and the concentration C can be represented as the sum of "average" and "fluctuating" components, i.e.,

$$V = \bar{u} + u', \tag{16}$$

$$C = \langle c \rangle + c', \tag{17}$$

where $\overline{(\ )}$ denotes the resolvable mean component, $(\ )'$ the remaining unresolvable fluctuating component, and $\langle\ \rangle$ the ensemble mean.   Then, substituting Eq. (16) and Eq. (17) in Eq. (15) and taking the ensemble average gives the following equation

$$\frac{\partial \langle c \rangle}{\partial t} = -\overline{u}.\nabla\langle c \rangle - \nabla.\langle c'u' \rangle + D\nabla^2\langle c \rangle + \langle Q \rangle, \tag{18}$$

in which it is assumed that $\langle \overline{u} \rangle = \overline{u}$ and $\langle u' \rangle = 0$. All Eulerian models solve some variation of the above equation, although they may differ in their initial conditions, spatial resolutions, boundary conditions, numerical techniques, simplifying assumptions, and turbulence parameterizations.

For example, K-theory dispersion models are the simplest of the Eulerian models and have been widely used. In K-theory dispersion models, the term $\nabla.\langle c'u' \rangle$ is simplified using the following approximation

$$\langle c'u' \rangle = -K\nabla\langle c \rangle, \tag{19}$$

where K is the turbulent diffusivity tensor. Furthermore, these models usually assume that the tensor K is diagonal (i.e., off-diagonal elements of K are zero), the molecular diffusion is negligible, and $\langle c \rangle$ represents the concentration of a nonreactive pollutant (i.e., $\langle Q \rangle = Q$). With these simplifications, K-theory dispersion models solve the following semi-empirical advection diffusion equation

$$\frac{\partial \langle c \rangle}{\partial t} = -\overline{u}.\nabla\langle c \rangle + \nabla. K\nabla\langle c \rangle + Q, \tag{20}$$

which can be integrated, if the inputs $\overline{u}$, K, and Q are provided, together with initial and boundary conditions for $\langle c \rangle$. Note that the velocity, $\overline{u}$, and turbulent diffusivity field, K, can be provided by a mesoscale meteorological model.

In general, Eulerian models suffer from several limitations. First, the treatment of advection in Eulerian models usually introduces artificial numerical diffusion and sometimes spurious oscillatory behavior, a problem when advecting non-negative physical quantities such as concentrations. Second, use of the gradient-transfer hypothesis limits these models to time scales much larger than the turbulence integral time scale and to pollutant spatial scales much larger than the turbulence integral spatial scales. Third, the pollutant mass being modeled must have a spatial extent at least equal to four or more horizontal and vertical grid increments in order that the gradients can be adequately defined and the advection phase errors minimized. Finally, these characteristics make point and line sources particularly difficult to treat in Eulerian models.

In spite of these shortcomings, Eulerian models are well suited to handle complicated physical processes such as reactive nonlinear chemistry (e.g., evaluation of O_3 impacts) and wet deposition. These models are also well suited to handle a large number of geographically fixed sources since the grid point calculations are independent of the number of sources.

11.2 Lagrangian dispersion models

Lagrangian models provide an alternative method for simulating atmospheric dispersion by defining fluid elements that follow the instantaneous flow. Emissions in Lagrangian models are represented by discrete elements (such as volumes and dimensionless points) transported by the mean wind and diffused by its turbulent components. The various specific approaches to Lagrangian modeling may differ in one or more of the following respects: the size, shape and other characteristics of the elements used to represent the emissions; the way in which diffusion processes are treated; the degree to which the individual elements respond to larger scale variability in the wind fields; and simplifications introduced to reduce calculation and computer memory requirements. Two of the most widely used Lagrangian dispersion model types are the Lagrangian puff model and the Lagrangian particle model.

11.2.1 Lagrangian puff models

Lagrangian puff models approximate the source emissions with a series of discrete "puffs". These models assume that each pollutant emission of duration, Δt, injects into the atmosphere a mass $\Delta m = Q\Delta t$, where Q is the time varying emission rate. The center of each puff containing the mass, Δm, is first advected according to the local time varying wind vector as follows:

$$x_i(t + \Delta t) = x_i(t) + \overline{u}_i(t)\Delta t , \quad \text{i=1,2,3} \qquad (21)$$

where x_i is the i-th component of the puff centroid location, and $\overline{u}_i$ is the i-th component of the resolvable or grid scale wind vector. The effect of atmospheric turbulence is then represented by increasing the physical dimension of the puffs (σ_x, σ_y, and σ_z) with travel time, travel distance, or at a rate appropriate to the turbulent state of atmosphere.

The distribution of material within the puffs can take almost any form, but for ease of calculation and for realism in the simulations, most implementations are limited to one of a few analytical functions (e.g., Gaussian). If, at time t, the center of a puff is located at $p(t) = (x, y, z)$, then the concentration due to that puff at the receptor $r = (x_r, y_r, z_r)$ can be computed using the following basic Gaussian puff formula

$$\Delta c = \frac{\Delta m}{(2\pi)^{3/2}\sigma_x\sigma_y\sigma_z}\exp\left[-\frac{1}{2}\left(\frac{x-x_r}{\sigma_x}\right)^2\right]\exp\left[-\frac{1}{2}\left(\frac{y-y_r}{\sigma_y}\right)^2\right]\exp\left[-\frac{1}{2}\left(\frac{z-z_r}{\sigma_z}\right)^2\right], \quad (22)$$

which is often expanded to incorporate reflection, deposition, and decay terms. The individual puffs, each of which contains the emissions for some specified time period, also overlap. Thus, the concentration at any point should be represented by the sum of the contribution from puffs near that point, rather than from a single puff.

Puff models are conceptually simple and this simplicity usually translates into relatively straightforward solutions to the mathematical and computer programming requirements. However, puff models usually ignore entrainment, cloud venting and other mixing effects with the ambient environment and they cannot accommodate non-linear chemistry. Also these models tend to produce symmetric concentration distributions, as they cannot easily deal with the effects of wind shear on scales smaller than the puff dimensions. Finally, treatment of multiple pollutant sources in puff models may quickly become very complicated.

11.2.2 Lagrangian particle models

The increased availability of inexpensive high performance computers has made it possible to replace Lagrangian puff models of atmospheric dispersion with more computationally intensive and more accurate Lagrangian particle dispersion models. These models are also known as Lagrangian stochastic models, Monte Carlo models, Markov chain models, or random walk models. Lagrangian particle dispersion models simulate the dispersion of pollutants in the atmosphere by means of a large ensemble of particles moving with pseudo-velocities. These pseudo-velocities simulate the effects of the two basic dispersion components: (1) transport due to the mean wind, and (2) diffusion due to the turbulent velocity fluctuations. The time dependent particle positions can then be computed from

$$x_i(t + \Delta t) = x_i(t) + \left[\overline{u}_i(t) + u_i'(t)\right]\Delta t, \quad i=1,2,3 \quad (23)$$

where x_i is the i-th component of the particle location, and $\overline{u}_i$ is the i-th component of the resolvable or grid scale wind vector, and u_i' is the i-th component of corresponding subgrid scale semi-random turbulent velocity fluctuation, which can be determined by means of the discrete Langevin equation

$$u_i'(t + \Delta t) = R_i(\Delta t)u_i'(t) + u_i''(t + \Delta t), \quad i=1,2,3 \tag{24}$$

where u_i'' is a purely random and normally distributed turbulent velocity component characterized by

$$\sigma_{u_i''} = \sigma_{u_i'}\left[1 - R_i^2(\Delta t)\right]^{1/2}, \quad i=1,2,3. \tag{25}$$

Here R_i is the i-th velocity component Lagrangian auto-correlation coefficient for time lag, and Δt can be related to Lagrangian turbulence time scales, T_L, by

$$R_i(\Delta t) = \exp\left[-\Delta t/T_L^i\right]. \quad i=1,2,3 \tag{26}$$

Generally, Lagrangian measurements of T_L are not available, but empirical relations have been proposed to estimate T_L from Eulerian meteorological measurements (e.g., Hanna, 1981). Equation (24) is written very generally in that turbulent velocity fluctuations are considered statistically independent. However, this assumption is in disagreement with wind fluctuation measurements that indicate the existence of non-zero cross-correlation terms. A number of investigators have treated the more general case, including these cross-correlation terms (Davis, 1983).

In Lagrangian particle models, each particle can be tagged by a mass of pollutant that can be either constant or time varying to allow loss of mass due to ground deposition and chemical decay phenomena. The spatial distribution of particle mass in the computational domain then allows the calculation of a three-dimensional mass concentration field. For example, the most straightforward method is to superimpose a three-dimensional concentration sampling grid on the calculation domain. The concentrations can then be computed by simply counting the number of particles in each sampling grid and accumulating their masses. In this method, the computed concentrations depend on the size of the sampling volume and on the number of particles used in the computation. To obtain a smooth and statistically steady concentration field, it is usually necessary to release a large number of particles (on the order of several thousand). The computational efficiency of the particle models can be significantly improved by the application of a kernel density estimator to calculate the concentration field from particle locations. The kernel method requires no imaginary sampling volumes and produces a smooth concentration distribution with a much smaller number of particles (Lorimer, 1986).

There are a number of disadvantages to Lagrangian particle models. First, they are not well suited for modeling the dispersion of nonlinear reacting pollutants. Second, the question of statistical sampling error must be addressed when estimating pollutant concentrations in areas of low particle density, sometimes requiring the release of a very large number of particles. Third, the most important and difficult task in particle modeling is to compute the Lagrangian statistics. Due to their Lagrangian nature, particle displacement should be computed using suitable Lagrangian measurements of the flow. Unfortunately, Lagrangian properties are difficult to measure and, therefore must be obtained from Eulerian measurements. Different methods have been proposed for relating Lagrangian turbulent statistics to the local Eulerian values, but none of them has been found totally satisfactory. Among many studies in this field, an acceptable statistical estimator has been presented by Hanna (1981). However, much research is still needed to extract the meteorological input from available meteorological measurements and to augment limited theoretical understanding of the turbulence processes needed to run these models.

In spite of these disadvantages, Lagrangian particle dispersion models have a number of advantages over other simulation techniques. First, Lagrangian particle models are practically free of restrictive physical assumptions, since they do not need the input of artificial stability classes, empirical sigma curves, or diffusion coefficients that are very difficult to measure. Instead, all uncertainties are combined into the correct determination of the pseudo-velocities. Second, a Lagrangian framework is more natural for modeling turbulent diffusion, which is a Lagrangian phenomenon. Third, Lagrangian particle models do not suffer from computational phase dispersion because no advection terms need to be considered in the mass conservation equations. Fourth, diffusion over small space and time scales, including emissions from multiple pollutant sources, can be treated easily by these models. Fifth, Lagrangian particle models are usually able to incorporate more turbulence properties than other simulation techniques due to their ability to utilize random turbulent components. Finally, Lagrangian particle dispersion models are flexible, conceptually simple, and computationally inexpensive when a reasonable number of particles are used. In principle, Lagrangian particle dispersion models can provide a degree of resolution and accuracy to complex flow solutions not obtainable by other simulation techniques, if the mean flow, Lagrangian time scales, and turbulent statistics are provided.

11.3 Current techniques to improve meteorology in dispersion models

Meteorological information such as wind fields and subgrid scale mixing and diffusion processes should be described in the mesoscale dispersion models in as detailed a manner as possible. There are two common approaches to provide this level of meteorological information: (1) the diagnostic approach, and (2) the prognostic approach. Important considerations in each approach are the data requirements, the computational requirements, and the input/output data flow requirements.

11.3.1 Diagnostic approach

The diagnostic approach tends to use wind field models or meteorological preprocessors that are based on the objective analysis of available meteorological observations.   The lowest level of diagnostic wind specification takes an observation at a single location and time, assumes that it is constant in space and time, and then performs a transport calculation in this simple flow field. A small increase in capability occurs when this type of system is modified in order to accept multiple observations at different times and then interpolates between those observations.   The next level of wind specification considers the blocking effect of the terrain.   This type of diagnostic wind specification is termed mass-consistent, because it builds a wind field from the observations that is forced to go around the terrain in such a fashion that the air mass is conserved. Given enough observations in the complex terrain, the mass consistent models can produce a reasonable flow field. Mass consistent models, however, cannot develop phenomena that do not yet exist, nor can they develop complex flow fields in the absence of data; the more complex the terrain, the more data they require.

An example of the diagnostic approach can be found in Pack *et al.* (1978) and Artz *et al.* (1985).   These studies diagnose the wind field by means of various objective analyses of twice-daily observations (every 12 hours) from the upper-air synoptic observing network. This network has an average station spacing of about 400 km in the United States.   Linear interpolation is then used to obtain the wind field between the synoptic observing times. These studies, however, have demonstrated that the existing conventional weather observing network cannot provide the necessary spatial and temporal resolution to incorporate mesoscale energy distributed over a variety of flow modes. The dispersion model predictions based on these synoptic-scale meteorological fields will likely be poor in many meteorological situations. This problem is especially serious within the planetary boundary layer and over complex terrain where the temporal and spatial inhomogeneities of the wind and turbulence fields are substantial. Also note that numerical interpolation from a coarse synoptic-scale meteorological field to a finer grid of a dispersion model does not necessarily increase resolution since no new information is provided. The resolution of the conventional weather observing network is the limiting factor.

11.3.2 Prognostic approach

An alternative approach to overcome the limitations of the existing weather observing network is to use meteorological fields predicted by prognostic mesoscale meteorological models. Although this approach is much more demanding conceptually and computationally, it provides physically consistent flow fields with better spatial resolution of the atmospheric flow information missed by the synoptic scale observing network and much better temporal resolution that preserves most of the diurnal and inertial energy. These forecast fields also include more variables, and can possess inter-variable thermodynamical and dynamical consistency among all of the fields. One widely used method to incorporate meteorological model predicted fields into dispersion models is to couple meteorological models with atmospheric dispersion models.   This coupling method uses three-dimensional gridded meteorological fields produced by prognostic mesoscale meteorological models at a selected time interval as input to dispersion models.

Many other investigators have also used the coupling method to study various mesoscale dispersion problems (e.g., McNider, 1981; Kao and Yamada, 1988; Pielke *et al.*, 1991; Boybeyi and Raman, 1995, Boybeyi *et al.*, 1995; Jin and Raman, 1996).   These studies simulated mesoscale dispersion for a wide spectrum of applications reasonably well. Although the coupling method has been sufficient for many studies, the major problem with this method is that atmospheric dispersion models can only use results from the meteorological models at a predetermined spatial and temporal resolution. The fundamental issue is how high a resolution is required for the atmospheric data in order to maintain the desired resolution of the atmospheric dispersion calculation. As the desired resolution increases, the amount of data that is required to be output increases as the cube. For example, increasing the horizontal grid resolution from 100 km down to 1 km increases the number of data elements by 10^4, and requires that the output occur 10^2 times as often in order to preserve accuracy - an increase in the input/output requirements of 10^6.

An alternative method to reduce the limitations of the coupling method is to embed a dispersion model within a mesoscale meteorological model (e.g., Boybeyi *et. al.*, 1994; Bacon *et. al.*, 1994). By embedding the dispersion model, the transport and diffusion algorithms have access to the meteorological data at the highest possible spatial and temporal resolution, thereby improving the dispersion model accuracy. The fundamental issue for prognostic systems using a coupled, as opposed to embedded dispersion methodology, remains the balance that must occur between the required resolution of the meteorological variables and the input/output load that places on the dispersion calculation.

However, it should be pointed out that even mesoscale meteorological models cannot provide perfect wind and turbulence fields necessary to run atmospheric dispersion models and that the magnitude of error in the prognostic fields usually increases with forecast time. Consider briefly the concepts of model scale and model resolution. Numerical modelers have found that the smallest temporal and spatial features that can be realistically treated in a numerical simulation are those $4\Delta x$ or $4\Delta t$ in size because of filtering and/or numerical smoothing (Pielke, 1984). Thus, even a mesoscale meteorological model with a typical horizontal grid spacing (say 50 km) will be unable to resolve some mesoscale flow features. A comprehensive discussion on better representation of meteorology in mesoscale models is given in Boybeyi and Bacon (1996).

12. Summary

Mesoscale atmospheric flows possess various mesoscale time and space scales resulting from the diurnal cycle, the atmospheric inertial mode, thermal and mechanical forcing due to mesoscale terrain inhomogeneities, and downscale energy transfer from the synoptic-scale to the mesoscale due to nonlinear flow interactions. Even for horizontally homogeneous flows over flat and uniform terrain, mesoscale frequencies such as the diurnal heating cycle and formation of a nocturnal low-level jet will usually be present.

The transport and diffusion of atmospheric pollutants can be affected by this wide range of flow scales through variations in the mean transport wind, differential advection due to vertical and horizontal wind shear, and vertical mixing. Therefore, these mesoscale space-time flow scales should be represented accurately when the mesoscale dispersion processes are studied. Otherwise, significant components of mesoscale transport and diffusion may be neglected. Finally, Embedding a dispersion model within a mesoscale meteorological model seems to have promising future.

References

Artz, R., Pielke, R. A. & Galloway, J. (1985) Comparison of the ARL/ATAD constant level and the NCAR isentropic trajectory analyses for selected case studies. *Atmospheric Environment*, **19**, 47-63.

Bacon, D. P., Boybeyi, Z., Dunn, T. J., Ho, Y-L., McCorcle, M. D., Peckham, S. E., Sarma, R. A., Young, S. & Zack, J. (1994) Development of the Operational Multiscale Environment model with Grid Adaptivity (OMEGA) Part I: Model description. *Proc. Tenth Conf. On Numerical Weather Prediction,* Portland, OR, American Meteorological Society, 538-540.

Batchelor, G. K. (1952) Diffusion in a field of homogeneous turbulence. II: The relative motion of particles. *Proc. Camb. Phil. Soc.*, **48**, 345-362.

Boybeyi, Z. & Raman, S. (1992) A three-dimensional numerical sensitivity study of convection over the Florida peninsula. *Boundary Layer Meteor.*, **60**, 325-359.

Boybeyi, Z., Bacon, D. P., Dunn, T. J., Ho, Y-L., McCorcle, M. D., Peckham, S. E., Sarma, R. A., Young, S. & Zack, J. (1994) Development of the Operational Multiscale Environment model with Grid Adaptivity (OMEGA) Part II: Simulation of local circulations. *Proc. Tenth Conf. On Numerical Weather Prediction,* Portland, OR, American Meteorological Society, 369-371.

Boybeyi, Z., Raman, S. & Paolo, Z. (1995) A numerical investigation of the possible role of local meteorology in the Bhopal gas accident. *Atmospheric Environment, 29, 479-496.*

Boybeyi, Z. & Raman, S. (1995) Simulation of elevated long-range plume transport using a mesoscale meteorological model. *Atmospheric Environment, 29, 2099-2111.*

Boybeyi, Z. & Bacon, D. P. (1996) The accurate representation of meteorology in mesoscale dispersion models. Chapter 4 of *Environmental Modeling-Vol. 3,* Edited by Paolo Zannetti, Computational Mechanics Publications, Southampton, Boston, 462 pp.

Brier, G. W. (1950) The statistical theory of turbulence and the problem of diffusion in the atmosphere. *J. Meteor.*, **7**, 283-290.

Davis, P. A. (1983) Markov chain simulations of vertical dispersion from elevated source into the neutral planetary boundary layer. *Boundary Layer Meteor.*, **26**, 355-376.

Draxler, R. R. (1988) The persistence of pollutants downwind of a point source following termination of the emission. *Boundary Layer Meteor.*, **42**, 43-53.

Hanna, S. R. (1981) Lagrangian and Eulerian time-scale relations in the daytime boundary layer. *J. Appl. Meteor.,* **20**, 242-249.

Jin, H. & Raman S. (1996) Dispersion of an elevated release in a coastal region, *J. Appl. Meteor.*, **35**, 1611-1624.

Kao, C.-Y. J. & Yamada, T. (1988) Use of the CAPTEX data for evaluations of a long-range transport numerical model with a four dimensional data assimilation technique. *Mon. Wea. Rev.*, **116**, 293-306.

Kolmogorov, A. N. (1941) The local scale structure of turbulence in incompressible viscous fluids for very large Reynolds numbers. *C. R. Acad. Sci. URSS*, **30**, 301-305.

Lilly, D. K. (1983) Stratified turbulence and mesoscale variability of the atmosphere. *J. Atmos. Sci.*, **40**, 749-761.

Lorimer, G. S. & Ross, D. G. (1986) The kernel method for air quality modeling-II. Comparison with analytic solutions. *Atmospheric Environment*, **20**, 1773-1780.

McNider, R. T. (1981) *Investigation of the impact of topographic circulations on the transport and dispersion of air pollutants.* Ph.D. dissertation, University of Virginia, Charlottesville, Virginia, 210 pp.

Moran, M. D. (1992) *Numerical modeling of mesoscale atmospheric dispersion.* Ph.D dissertation, Colorado State University, Fort Collins, Colorado, 758 pp.

Orlanski, I. (1975) A rational subdivision of scales for atmospheric processes. *Bull. Amer. Meteor. Soc.*, **56**, 527-530.

Pack, D. H., Ferber, G. J., Heffter, J. L., Telegadas, K., Angell, J. K., Hoecker, W. H. & Machta, L. (1978) Meteorology of long range transport. *Atmospheric Environment*, **12**, 425-444.

Pasquill, F. (1974) Limitations and prospects in the estimation of dispersion of pollution on a regional scale. *Adv. Geophys.*, **18B**, 1-13.

Pielke, R. A. (1984) *Mesoscale Meteorological Modeling.* Academic Press, Orlando, Florida, 612 pp.

Pielke, R. A., Lyons, W. A., McNider, R. T., Moran, M. D., Moon, D. A., Stocker, R. A., Walko, R. L. & Uliasz, M. (1991) Regional and mesoscale meteorological modeling as applied to air quality studies. In *Air Pollution Modeling and Its Application* VIII, van Dop, H. And D. G. Steyn, Editors, Plenum Press, New York, 259-289.

Randerson, D (1984) Atmospheric Science and Power Production, USDOE Report TIC-27601. Available from National Technical Information Service, Springfield, VA.

Slade, D. H. (1968) *Meteorology and Atomic Energy*, USAEC Report TID-24190. Available from National Technical Information Service, Springfield, VA

Stull, R. B. (1988) *An Introduction to Boundary Layer Meteorology*. Kluwer Academic Publisher, Massachusetts 666 pp.

Taylor, G. I. (1921) Diffusion by continuous movements. *Proc. R. Soc (London) Ser. A*, **164**, 196-212.

Zannetti, P. (1990) *Air Pollution Modeling*, Van Nostrand Reinhold, New York, 444 pp.

Chapter 5
Leaky containment vessels of air:
A Lagrangian-mean approach to
the stratospheric tracer transport

Noboru Nakamura
Department of the Geophysical Sciences,
The University of Chicago, Chicago, IL 60637, USA

Abstract

In this article, we shall review the tracer transport in the Earth's stratosphere from a Lagrangian-mean perspective. A new diagnostic formalism is introduced in which the tracer's distribution and motion are measured with respect to mass enclosed by the tracer's isosurface in the isentropic layer. This naturally characterizes containment vessels of air and their "leakiness". The relative motion of the tracer and mass reflects nonconservative processes including isentropic diffusion and diabatic heating to which the present method provides particularly clean interpretations. Concentrated tracer gradients (e.g. polar vortex edge) and their permeability are also resolved more precisely than the conventional methods. Application of the diagnostic is demonstrated using the nitrous oxide field simulated by the Geophysical Fluid Dynamics Laboratory's SKYHI general circulation model wherein the natural variabilities of the tracer transport in the stratosphere are illustrated.

1 Introduction

The main emphasis of this volume up to this point has been on the atmospheric boundary layer (ABL) as a source region of the anthropogenic pollutants. The chemicals escaping ABL into the free troposphere will eventually reach the stratosphere provided that their life time is sufficiently long. While there is much to be learned about the transport between ABL and the base of the stratosphere, the transport processes within the stratosphere is one area of atmospheric science that has seen a considerable progress in recent years and deserves a room for discussion in its own right. The following material will be largely drawn from the author's own work but also reflect the gist of the latest developments in the field to be representative. The readers interested in the exchange of substances between the troposphere and stratosphere shall also benefit from an excellent review by Holton *et al.* (1995).

The Earth's stratosphere—particularly the lower part of it (15-30km in altitude)—is unique in that photochemistry, radiation and large-scale transport operate at comparable time scales to sustain an intricate climate balance (Andrews *et al.* 1987). Of particular importance is the behavior of the ozone layer that shields life on Earth against the harmful ultraviolet (UV) solar radiation. The discovery of the Antarctic ozone hole in the mid 80s was especially alarming in that it demonstrated the vulnerability of the photochemical equilibrium of the ozone layer to the anthropogenic emission of gases. While it has been determined that catalytic photochemistry is at the heart of the ozone hole, the large-scale transport plays a crucial role of regulating the air temperature and abundance of the reactants that set the stage for the chemistry (e.g. McIntyre 1989). Lately the stratospheric research has been propelled by the increasing availability of remote sensing and airborne measurements. To dynamicists, the large crop of new constituents data, such as those provided by the Upper Atmosphere Research Satellite (UARS), presents a fertile ground for investigating the transport processes and validating their numerical simulations.

The stratosphere is characterized by a very slow vertical mixing and disproportionately fast horizontal mixing, and in this sense is fundamentally different from ABL. The high stratification suppresses the vertical excursion of air, and the motion is largely confined to within the isentropic surface on a short time scale. The systematic vertical transport in the stratosphere is mostly associated with radiative heating, small-scale turbulence caused by occasional gravity wave breaking and molecular diffusion. Even the radiative process, the most efficient of these, has a characteristic time scale as long as a few months (Hitchman and Leovy 1986). In comparison, the horizontal excursion of air can be very fast, particularly in winter. For example, the extratropical stratosphere in the Northern Hemisphere winter sustains a most vigorous lateral mixing observed in any part of the atmosphere in any season. This hemispheric "stirring", whose time scale is of the order of a few days, is primarily associated with the planetary waves excited by the westerly flow impinging the large-scale surface orography (Andrews *et al.* 1987).

As we shall see in the next section, the horizontal distribution of the trace constituents in the winter stratosphere exhibits striking inhomogeneity *despite* the fast stirring. The degree of inhomogeneity is such that it cannot be explained by the source-sink distribution or by the differential welling associated with the variation in the heating rates. This suggests that horizontal mixing itself is spatially inhomogeneous. It is now believed that the stratosphere consists of a few well-mixed containment vessels that are dynamically separated by *mixing barriers*, at least during certain periods of the year. For example, the Antarctic air in the early spring seems well isolated by the remnant of the *polar night jet*, a concentrated westerly flow surrounding the winter pole, allowing the low-temperature catalytic photochemistry inside it to destroy ozone with little dilution (McIntyre 1989, Schoeberl *et al.* 1992, Bowman 1993). Plumb (1996) also discusses a similar dynamical isolation of the tropical stratosphere.

An analogy may be drawn for ABL in the presence of temperature inversion: the well-mixed air is confined to ABL to the extent the inversion acts as a "barrier" to the vertical mixing. However, a critical difference between ABL and the stratosphere is that the containment vessel for the latter can be defined only in a Lagrangian sense. In contrast to ABL which by definition rests near the ground, the hypothetical container embedded in the stratosphere moves with the large-scale wind and hence is generally unsteady. Indeed the air surrounded by the polar night jet "wobbles" significantly in the Northern Hemisphere winter (Juckes and McIntyre 1987; Polvani and Plumb 1992). While this reversible material undulation to a large degree characterizes the air motion in the stratosphere, some air parcels do escape the containment and hence make the vessel "leaky", so to speak. This complicates the quantification of transport and mixing based on the containment vessel model: one must distinguish the drift and deformation of the container itself from the entrainment and detrainment through the moving wall of the container, as schematized in Fig.1. Each of these kinematic properties of the containment vessel has distinct implications for the radiative and photochemical processes embedded within.

The drift of the container can affect the inside solar irradiance and temperature because radiation is a strong function of latitude and height. This in turn modifies the abundance of chemical substances in the container by perturbing

SOLAR IRRADIANCE

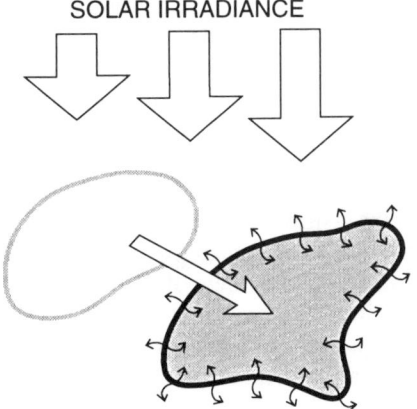

Figure 1 Schematic view of a leaky containment vessel of air. The thin white arrow indicates the horizontal drift of the container, whereas the double headed wavy arrows indicate the leakage of substances through the bounding wall of the container. See text for details.

the reaction rates. For example, whether the air containing reactants resides in the polar night region or in the sunlit region makes much difference in the UV-sensitive photochemistry (Solomon *et al.* 1993). The drift of the container to a latitude with a different radiative equilibrium temperature also modifies the heating-cooling rates within it (Fels 1985).

Meanwhile, the entrainment and detrainment processes through the wall of the container also influence the inside chemical abundance and temperature. As described earlier, the dynamical stirring associated with the planetary wave breaking makes the air that resides in the stratospheric polar vortex, the region bounded by the polar night jet, less isolated (Juckes and McIntyre 1987, Dritschel 1988, Baldwin and Holton 1988, Polvani and Plumb 1992, Newman *et al.* 1993, Waugh *et al.* 1994). Temperatures in the dynamically more active Arctic vortex are thus higher than in the Antarctic, preventing a major ozone hole from developing (so far).

The goal of this article is to quantify this second aspect of transport, that is, the *leakiness* of the container. As Fig.1 indicates, the degree of containment (or leakage) can be best quantified if one altogether abandons reference to the geographically fixed coordinates and redefine transport relative to the container wall wherever it is at. We shall develop a rigorous mathematical formalism to achieve this goal. For a maximal conceptual clarity, emphasis will be placed on the nature of passive tracers simulated by various numerical models. Retreat from photochemistry and radiative transfer is simply a means to keep our focus on transport, and should not be considered as a sign of irrelevance. In fact, as the foregoing discussion suggests, much of the photochemical and radiative drive (e.g. solar irradiance) is affected by the other half of transport, the drift of the container, and hence lies beyond the scope of this article.

Help of computational fluid mechanics is indispensable even for the purpose of developing diagnostics. The essence of the transport-mixing problem is the transfer of information between spatial scales that are decades apart, which is difficult to address with the available stratospheric measurements whose resolution and coverage are limited. We shall therefore utilize a hierarchy of numerical models, from a simple kinematic model to a fully configured general circulation model (GCM), as testbeds for our diagnostic.

Organization of the material is as follows. In the next section, the nature of the problem will be illustrated using the distribution of the stratospheric nitrous oxide (N_2O) simulated by the Geophysical Fluid Dynamics Laboratory (GFDL) SKYHI GCM. We shall also present a brief overview of the traditional diagnostic formalisms for the tracer transport, and discuss why they are not very satisfactory for quantifying the concept of a leaky containment vessel. Then we shall introduce a new diagnostic strategy, with remarks on some important preceding developments that formed a basis for the present work. In section 3, we demonstrate the development and application of the new diagnostic in the kinematics of two-dimensional mixing on the isentropic surfaces. There we shall see how chaotic advection induces fast mixing, and how spatially varying stirring leads to a formation of extreme tracer inhomogeneities such as those found at the edge of the polar vortex. We shall also quantify the transport barriers and permeability, which are needed to solidify the notion of a "leaky container". In section 4, we shall generalize the formalism to include cross-isentropic transport by diabatic processes. Utility of the method is then demonstrated in the diagnosis of the SKYHI GCM output. In particular, how the containers and their leakiness evolve with seasons, and how they vary between the two hemispheres and over different years will be illustrated. In section 5, a formal connection between the new and traditional formalisms is noted in terms of the nonacceleration theorem. We will conclude with remarks on the challenges that we face in incorporating photochemistry and radiation to the present formalism.

While the method is constructed with an immediate application to the stratospheric transport in mind, the reader should be reminded that the conceptual and mathematical development to be presented below is completely general and thus pertains to a wider spectrum of transport problems in the atmosphere and ocean. Other potential applications will be touched on in section 6.

2 Characterizing the stratospheric transport

2.1 Nature of the problem

The distribution of the trace constituents in the stratosphere is commonly presented as a zonally averaged field. However, this traditional method is not amenable to the concept of leaky containment vessel. This is easy to see, for example, in a snapshot of stratospheric tracer distribution in the Northern Hemisphere winter. Figure 2 shows a hemispheric distribution of the nitrous oxide (N_2O) mixing ratio in the lower stratosphere, taken from the GFDL SKYHI GCM simulation. [For a description of the model, see Fels et al. (1980) and section 4.3.] Here the use of a GCM output is purely aesthetic: satellite measurements of the stratospheric N_2O in the northern winter exhibit similar traits except at much lower resolution (e.g. Ruth et al. 1994, Roche et al. 1996).

Nitrous oxide is a chemically inert tracer in the lower stratosphere, and hence its distribution is governed primarily by transport. The N_2O mixing ratio generally decreases from the equator to the poles. This is a result of differential welling associated with the heating distribution [see section 4]. However, the horizontal gradients of the tracer are very inhomogeneous. If we are to distinguish air masses according to the N_2O mixing ratio, three distinct regions can

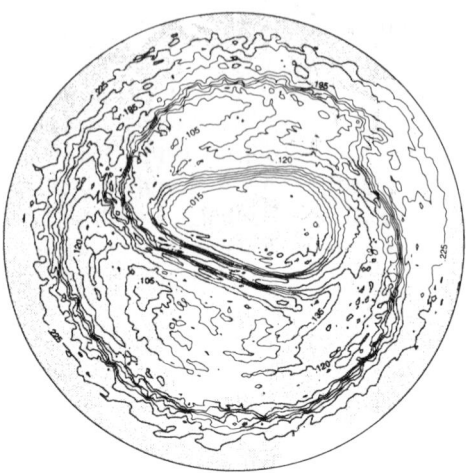

Figure 2 Azimuthal equidistant plot of the Northern Hemisphere N_2O mixing ratio on 650K isentropic surface at 1200 UTC 2 February 1984 (model year), simulated by GFDL SKYHI GCM (horizontal resolution: 1.2x1.0 degrees). The center and outer edge of the figure correspond to the north pole and the equator, respectively. Contour interval is .015 ppm. [From Nakamura (1995). Copyright: American Meteorological Society.]

be easily identified: the extremely N_2O-poor main (polar) vortex, the midlatitude *surf zone* with relatively weak tracer gradients, and the tropical reservoir with large N_2O abundance (McIntyre and Palmer 1983, 1984; Randel *et al.* 1993). These regions are separated by concentrated horizontal gradients (edges) of the tracer. These edge regions are found to be remarkably coherent throughout the Northern Hemisphere winter in the SKYHI simulation, although their shapes are highly transient and irregular due to the underlying activity of the planetary waves. Aircraft measurements of various tracers in the lower stratosphere provide a circumstantial evidence that the tracer edges do exist in the real atmosphere (Schoeberl *et al.* 1992, Strahan and Mahlman 1994a). It is thus natural to think of these edge regions as a boundary of distinct air masses, and thereby use them to define "container walls" (Tao and Tuck 1994, Nash *et al.* 1996).

The problems with the zonally averaged models are, first of all, they cannot fully resolve the tracer edges that are displaced from a latitude circle: the zonal averaging smears the meandering edges, creating a false sense of diffuseness. Inability to resolve the edges makes it even harder to estimate how much exchange of substance takes place across the edges (or how leaky the containers are). For example, for the notion of containment vessel to be valid, the bounding edge should exhibit a limited permeability and hence act as a transport barrier. Are the edges indeed signature of transport barriers as widely quoted, and if so, how can one quantify the strength of a barrier? To answer these questions, the flux of substances must evidently be measured with respect to the meandering edges, not to the geographically fixed latitude. Such exchange is inherently associated with nonconservative processes: as long as the tracer is conservative, its isosurfaces are a material surface and hence no air parcel can cross them. Much of this nonconservation is due to diffusive mixing and thus of irreversible nature. The Eulerian zonal-mean models cannot discriminate true irreversible transport from reversible undulation (by nonbreaking planetary waves, for example). Irreversibility in the zonal-mean transport is usually parameterized by the eddy diffusivity, based on the assumption that the scale of the mean gradients are much larger than that of the eddy (McIntyre 1992). This assumption approximately holds for small amplitude eddies (Plumb and Mahlman 1987) but fails precisely in the situations like Fig.2 wherein the scale of the eddy is comparable to, or larger than, that of the mean gradients. The dubious validity of eddy diffusivity concept prompts us to seek alternative diagnostic for horizontal mixing.

2.2 Extant Lagrangian diagnostics

To overcome the shortcomings of the Eulerian zonal mean models, it is necessary to invoke some Lagrangian-mean perspectives. Among the most established alternative formalisms currently available are the generalized Lagrangian mean (GLM) and its Eulerian sibling, the transformed Eulerian mean (TEM) (Andrews and McIntyre 1976, 1978; Boyd 1976). In essence, these formalisms include the latitudinal Stokes drift of the material contours as part of the mean transport velocity, while defining eddy transport with respect to the drift-corrected latitude. This transformation provides a deeper insight on the wave-mean flow interaction, and the obtained residual mean meridional circulation is a Lagrangian mass flow (essentially the diabatically driven Brewer-Dobson circulation; Dunkerton 1978) which is more pertinent to the tracer transport than the Eulerian mean circulation. Nevertheless, the mean Stokes drift does not take into account the change in the shape of a container: The mean fields of GLM and TEM are in practice still defined to be zonally symmetric, and hence unable to capture the wavy tracer edges at a given instant like those in Fig.2. As far as the characterization of the tracer inhomogeneity is concerned, these formalisms offer little improvement over the Eulerian zonal mean approach.

Another brand of Lagrangian perspective involves advecting a number of particle points or geometrical objects by the known wind. This line of work dates back to Welander (1955), and a large quantity of literature exists today on the particle trajectory analyses in two- and three-dimensions (e.g. Kida 1983, Pierce and Fairlie 1993, Pierrehumbert and Yang 1993, Bowman 1993, Pierce et al. 1994, Manney et al. 1994, Sutton et al. 1994, Dahlberg and Bowman 1994, Eluszkiewicz et al. 1995). Some recent studies explore new diagnostic ideas imported from the theory of dynamical systems (e.g. Lyapunov exponents and probability distribution functions; e.g. Weiss 1991, Pierrehumbert

1991ab, Yang 1993). Unlike the Eulerian approach, the particle tracer methods characterize transport, mixing and barriers thereto by keeping track of the position of air parcels represented by each particle. A variant of this method is the contour advection, in which the number of particles is continually adjusted so as to preserve the particle density along each material contour (Waugh and Plumb 1994, Norton 1994). The contour advection method captures filamentation of the material contours very precisely, to much finer scales than those resolved by the high-resolution GCMs.

The main difficulty with the particle advection methods lies in the fact that the motion of air parcel is not identical with that of tracer. Due to chaotic stretching and scrambling, the material contours in the atmosphere develop fine scale filaments in short time (Pierrehumbert and Yang 1993). The embedded tracer would then become subject to diffusion, resulting in the dissipation of the tracer filaments, while material contours can be further stretched. This is precisely how tracer is irreversibly transported across the material contours (or air parcel transported across the tracer contours). The particle methods address scrambling of material contours but not the exchange of tracer molecules across them (the *ad hoc* contour surgery associated with contour advection is a localized procedure and quite distinct from diffusion; Waugh and Plumb 1994). Of course one can define a sense of diffusion from a coarse-grained statistics of particles, but such diffusivity would depend arbitrarily on the chosen scale of the grain.

Lack of a useful flux-gradient relationship is another problem with the particle methods. Because of this, they do not address how gradients drive the transport, and the mixing properties must be addressed in terms of an initial value problem. However, one obtains a visual impression from Fig.2 that *mixedness* and *stirredness* of a tracer can be diagnosed instantaneously to within the resolution of data without solving an initial value problem.

2.3 Modified Lagrangian mean diagnostic

In view of the above, an altogether different approach seems to be needed to quantify the entrainment/detrainment processes into and out of containment vessels of irregular shape. The following development traces its root to the modified Lagrangian mean (MLM) concept of McIntyre (1980), who proposed the use of potential vorticity and potential temperature as Lagrangian labeling coordinates. Since then, the theme has been explored by several authors in various contexts (e.g. Dunkerton *et al.* 1981, McIntyre 1982, Butchart and Remsberg 1986, Schoeberl and Lait 1991, Schoeberl *et al.* 1992, Manney *et al.* 1994, Norton 1994, Lary *et al.* 1995). In recent papers the author has attempted to generalize, and somewhat unify, these isolated applications into a formal transport model (Nakamura 1995, 1996). This formulation has descended most directly from Butchart and Remsberg (1986) who studied the effects of diabatic heating in the stratosphere using the area enclosed by potential vorticity contours as a diagnostic.

The proposed MLM diagnostic characterizes a containment vessel in terms of the air mass enclosed by the tracer isosurfaces, and its leakiness in terms of the mass transport relative to tracer isosurfaces. This can also be interpreted as the *tracer transport with respect to material surfaces*. Thus the formalism allows for a compact representation of the two transport processes that are otherwise treated separately. As pointed out earlier, the nonconservative processes such as diffusion and diabatic heating are crucial for this type of transport: If the tracer were perfectly conserved following the motion of air, the tracer isosurfaces would be a material surface. Hence there would be no transport of mass across the tracer isosurfaces regardless of their position and shape. The relative transport of mass and tracer is precisely the point missed by the particle advection methods.

Since we are primarily concerned with the amount of mass enclosed by the tracer isosurface and its nonconservative leakage and not the position of the container itself, it suffices to examine simple geometrical attributes of the container without a reference to the Eulerian coordinates. Decomposition of the field into *wave* and *mean* is unnecessary, and hence much complication associated with the full nonlinearity and zonal asymmetry, common among the conventional methods, is avoided.

Of all the possible nonconservative effects, we give a special attention to horizontal mixing and diabatic effects, the two major drives for the chemically inert tracers in the stratosphere. We first demonstrate the development of MLM formalism for the horizontal mixing, and then later include the diabatic effects to complete the tracer kinematics in the vertical cross section.

3 Isentropic mixing

Absence of organized convection makes the vertical mixing in the stratosphere much slower than in the troposphere. The large static stability inhibits the adiabatic vertical excursion of air parcels. However, there is less constraint on the quasi-horizontal, isentropic excursion of air parcels. Indeed, the small inertia of the stratospheric air renders it very responsive to even a modest amount of dynamical forcing. The stratospheric sudden warming in the Northern Hemisphere winter is perhaps the most drastic illustration of this (McIntyre 1982), but even in the absence of a sudden warming, action of the planetary waves in the Northern Hemisphere winter is commonly observed in the highly zonally asymmetric contours of a quasi-conservative tracer, as Fig.2 clearly illustrates.

The turnover time scale associated with these finite-amplitude planetary waves (eddies) is of the order of days, which is much shorter than the time scale associated with the cross-isentropic (diabatic) transport (order of months; Hitchman and Leovy 1986, Gille *et al.* 1987). Hence the essence of the eddy transport should be well modeled by the two-dimensional kinematics on the isentropic surfaces.

3.1 Transport equation in the area coordinate

In this section, we develop an MLM diagnostic to examine the nature of the horizontal mixing in isolation. Much of the development below follows Nakamura (1996). The entry point of our inquiry is the simple advection-diffusion equation for the tracer mixing ratio q

$$\frac{\partial}{\partial t} q + J(\psi, q) = D \nabla^2 q. \tag{1}$$

Here ψ, J, and ∇^2 denote streamfunction, Jacobian and Laplacian, respectively, defined on the isentropic surface. Use of the streamfunction assumes that the stratospheric winds are nearly nondivergent in a short time scale, which is justifiable through scale analysis. The constant diffusion coefficient D represents either molecular, turbulent or numerical subgrid diffusion: in other words *microscale* diffusion. Other terms affecting the tracer mixing ratio, such as photochemical sources-sinks and diabatic transport, would have appeared as a source term on the right-hand side of eqn (1) but are altogether neglected. These neglected slower processes would regulate the overall gradient structure of the tracer, but not the fast kinematics of the tracer contours.

To define the containment vessel and quantify transport through its wall, we start, instead of directly manipulating eqn (1), by associating each tracer contour and the area (mass) of air at a given time:

$$\textit{area of the region where } q^* \leq q \quad = \iint_{q^* \leq q} dA \equiv A(q,t), \tag{2}$$

where q^* is a descriptor of the tracer contour. If the inequality in eqn (2) is reversed, then $A(q,t)$ denotes the area that resides on the other side of the contour $q^* = q$. Relation between area A and tracer q in eqn (2) is one to one (unless q is everywhere uniform) because

$$A\left(q+\delta q,\, t\right) - A\left(q,\, t\right) \;=\; \int\!\!\int_{q\le q^{*}\le q+\delta q} dA \;>\; 0 \quad \text{for} \quad \delta q > 0. \tag{3}$$

This one-to-one relation is intuitive in a stratospheric situation like Fig.2, where the tracer contours form quasi-concentric loops. Yet the relation is actually entirely independent of the geometry of the contours: the contours can be contorted, filamented, or even broken (multiply connected). The area enclosed by the broken contours is lumped together with the main area. This does not constitute a conceptual problem, since the air peeled away from the main body of air always stands a chance of being re-merged to the main body until its Lagrangian identity is lost nonconservatively. Although the tracer contours can be terminated at the rigid boundaries, in this article we restrict our attention to the self-closed (periodic) contours.

An immediate consequence of this universal one-to-one relation is that one can map the tracer q using A as the horizontal coordinate, $q(A,t)$, thereby reducing the spatial dimension by one. Since this representation does not depend on the shape of the contour, much technical difficulty associated with the full nonlinearity of finite-amplitude eddy can be avoided. In fact, zonally asymmetric eddies in the conventional sense are now absorbed as part of the *mean* state defined by $q(A,t)$.

At this point it is convenient to introduce an area integral operator $\mathcal{A}$:

$$\mathcal{A}(\,\cdot\,) \;\equiv\; \int\!\!\int_{q^{*}\le q} (\,\cdot\,)\, dA \;=\; \int_{q_{min}}^{q} dq^{*} \oint_{q^{*}} \frac{(\,\cdot\,)}{\left|\nabla q\right|}\, dl. \tag{4}$$

The last step of eqn (4) essentially divides the area into a series of "rings" defined by the closed q-contours (Fig.3). When a ring is multiply connected,

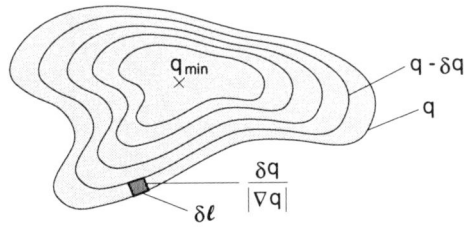

Figure 3 Illustration of the last step of eqn (4). Procedure is similar when the contour level *increases* inwardly.

summation is taken for all "islands" to evaluate the circuit integral. This final expression is necessary to complete the rest of the formalism, but not actually used for the numerical evaluation of the area integral. Useful corollaries of eqn (4) are

$$A(q,t) \;=\; c\mathcal{A}(1) \;=\; \int_{q_{min}}^{q} dq^* \oint_{q^*} \frac{dl}{|\nabla q|}, \tag{5}$$

$$\frac{\partial}{\partial q} c\mathcal{A}(\,\cdot\,) \;=\; \oint_{q} \frac{(\,\cdot\,)}{|\nabla q|} dl. \tag{6}$$

Using the area integral operator, average of any scalar around the q-contour is defined as

$$(\,\overset{\wedge}{\cdot}\,) \;\equiv\; \frac{\partial}{\partial A} c\mathcal{A}(\,\cdot\,), \tag{7}$$

while it is easy to show $\hat{q} = q$. Notice the mean quantity defined by eqn (7) is *nonlocal*, that is, the evaluation of the contour average requires knowledge of the scalar field outside the contour (except for q).

To prepare ourselves to derive the governing equations for $A(q,t)$ and $q(A,t)$, we further take note of an identity

$$\frac{\partial}{\partial t} \iint_{q^* \le q} (\,\cdot\,) \, dA \;=\; -\frac{\partial}{\partial q} \iint_{q^* \le q} (\,\cdot\,) \frac{\partial q}{\partial t} dA \;+\; \iint_{q^* \le q} \frac{\partial (\,\cdot\,)}{\partial t} dA, \tag{8}$$

where $\partial/\partial t$ on the left hand side denotes time derivative with q fixed, while those on the right hand side denote time derivative at fixed Eulerian coordinates. A proof of eqn (8) is found in the appendix. Using eqns (8), (4), and (5),

$$\frac{\partial}{\partial t} A(q,t) \;=\; -\frac{\partial}{\partial q} \iint_{q^* \le q} \frac{\partial q}{\partial t} dA$$

$$=\; -\frac{\partial}{\partial q} c\mathcal{A}\big(D\nabla^2 q - J(\psi,q)\big). \qquad \leftarrow Eqn\ (1) \tag{9}$$

However, since nondivergent flow is area-preserving,

$$\frac{\partial}{\partial q}\mathcal{A}\left(J(\psi,q)\right) \;=\; \oint_q \frac{J(\psi,q)}{|\nabla q|}\,dl \qquad\qquad \leftarrow Eqn\ (6)$$

$$=\; \oint_q \frac{(\mathbf{k}\times\nabla\psi)\cdot\nabla q}{|\nabla q|}\,dl \;=\; \oint_q (\mathbf{k}\times\nabla\psi)\cdot\mathbf{n}\,dl$$

$$=\; \iint_{q^*\le q} \nabla\cdot(\mathbf{k}\times\nabla\psi)\,dA \qquad \leftarrow divergence\ theorem$$

$$=\; 0,$$

and hence eqn (9) reduces to

$$\frac{\partial}{\partial t}A(q,t) \;=\; -\frac{\partial}{\partial q}\mathcal{A}\left(D\nabla^2 q\right). \tag{10}$$

Equation (10) expresses the mass continuity in the tracer (q) coordinate. The right hand side denotes the mass flux convergence into the container enclosed by the q-contour due to diffusion. It vanishes if $D=0$ because then the tracer is conserved following the motion of air, hence their contours become a material line. The governing equation for $q(A,t)$ is derived by inverting eqn (10):

$$\frac{\partial}{\partial t}q(A,t) \;=\; -\frac{\partial}{\partial t}A(q,t)\times\frac{\partial q}{\partial A}$$

$$=\; \frac{\partial}{\partial A}\mathcal{A}\left(D\nabla^2 q\right). \qquad \leftarrow Eqn\ (10) \tag{11}$$

However, since

$$\mathcal{A}\left(D\nabla^2 q\right) \;=\; D\iint_{q^*\le q} \nabla\cdot\nabla q\,dA$$

$$= D \oint_{q^{\cdot}=q} \nabla q \cdot \frac{\nabla q}{|\nabla q|} \, dl \quad \leftarrow \textit{divergence theorem}$$

$$= D \frac{\partial}{\partial q} c\mathcal{A}\left(|\nabla q|^2 \right) \quad \leftarrow \textit{Eqn (6)}$$

$$= D \left(\frac{\partial q}{\partial A} \right)^{-1} \frac{\partial}{\partial A} c\mathcal{A}\left(|\nabla q|^2 \right)$$

$$= D \left(\frac{\partial q}{\partial A} \right)^{-1} |\hat{\nabla q}|^2, \quad \leftarrow \textit{Eqn (7)} \qquad (12)$$

eqn (11) can be further rewritten as

$$\frac{\partial}{\partial t} q(A,t) = D \frac{\partial}{\partial A}\left(L_e^2 \frac{\partial q}{\partial A} \right), \qquad (13)$$

$$L_e^2(A,t) = |\hat{\nabla q}|^2 \left(\partial q/\partial A \right)^{-2}. \qquad (14)$$

Here using area A as the horizontal coordinate, we have reduced eqn (1) to a one dimensional diffusion equation (13), and thereby introduced a free parameter L_e^2 defined by eqn (14). We call L_e the equivalent length of the q-contour.

3.2 The equivalent length

Equivalent length defined by eqn (14) measures, in effect, the perimeter of the containment vessel. It reduces to the actual length of the q-contour (in the sense of line integral) when $|\nabla q|$ is constant around the contour, since in that case $L_e = |\nabla q| / |\partial q/\partial A|$ from eqn (5), but in general they are slightly different (Fig.4). That L_e^2 is defined in terms of area integrals and their derivatives, instead of a line integral, makes its numerical evaluation easier: Implementing line integrals over contorted contours can be time-consuming and taking inventory of broken contours requires an involved programming.

   Equivalent length links advection and diffusion in eqn (13). Although microscale diffusion (nonzero D) is necessary to bring about the cross-mass transport of tracer, it alone achieves little mixing because of the small mixing length. In order for the tracer to be modified over large area in short time,

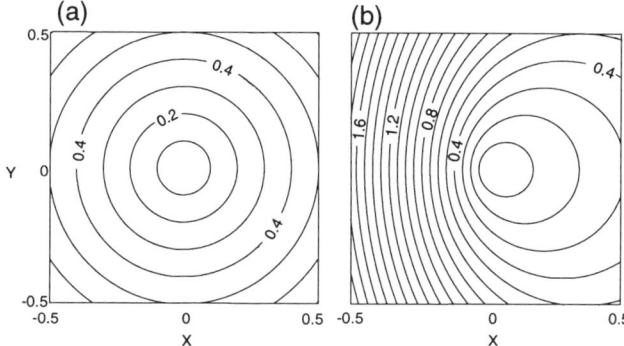

Figure 4 The above two frames show the isopleths of tracer q whose distribution is given by $q = (-\alpha x + \sqrt{x^2 + (1-\alpha^2)y^2}\,)/(1-\alpha^2)$, where the two cases correspond to (a) $\alpha = 0$, (b) $\alpha = 0.7$. Notice that the isopleths are a perfect circle in both (a) and (b), and the perimeter of the isopleths holding the same value is identical between the two frames. However, because of the asymmetrical distribution, *equivalent length* L_e (normalized by radius) is greater for (b): i.e. $L_e / 2\pi R = 1.000$ for (a) whereas $L_e / 2\pi R = 1.183$ for (b), R being the radius of the isopleth. Thus, (b) is more susceptible to diffusion.

equivalent length must be stretched substantially. Stretching of equivalent length is precisely the role of chaotic advection. In the low viscosity limit, it is achieved by creating a large power of $|\nabla q|^2$ in small scales by cascading the tracer variability from larger scales (Ottino 1989, Pierrehumbert 1991a). Thus, chaotic advection (stirring) promotes mixing by enhancing the available interface across which diffusion operates. Here it is important to differentiate stirring from mixing (Wiggins 1988). While L_e^2 measures stirredness or scrambledness of the tracer contours, the areal gradient $\partial q / \partial A$ measures mixedness of the tracer.

Of course diffusion can in turn affect equivalent length. Smearing of small-scale wriggles in the contours, for example, may limit the growth of the length by reducing $|\nabla q|^2$. On the other hand, as the tracer becomes homogenized, diminished areal gradient $\partial q / \partial A$ may tend to enhance L_e^2 [eqn (14)]. Thus the behavior of L_e^2 in the viscous limit is not at all trivial. Since we cannot express $|\nabla q|^2$ in terms of the area-mean quantities alone, prediction of L_e^2 remains a difficult closure problem. [It has been brought to the author's attention that Rhines and Young (1983) derived eqn (13) for limiting cases in which tracer contours are nearly parallel to streamlines (their eqn 4.12). Their result suggests L_e^2 can be modeled using streamfunction and circulation in this limit.] Regardless of how it is determined, once known, L_e^2 serves as a useful diagnostic for the efficiency of irreversible transport.

In eqn (13), DL_e^2 can be thought of as a Lagrangian equivalent of the eddy diffusion coefficient (K_{yy}) in the zonal mean formalisms. Unlike K_{yy}, however, validity of DL_e^2 is sound both physically and mathematically. Arbitrary time averaging is unnecessary to make this coefficient positive—it is defined instantaneously and positive by definition. Furthermore, L_e^2 is determined solely by the geometrical attributes of the tracer contours, and thus calculable without using winds. This is particularly attractive for the stratospheric applications where wind data are not readily available observationally.

3.3 Permeability and barriers

How equivalent length regulates the mixing of tracer molecules can be best understood by forming a "random walk" problem from eqn (13). Discretize eqn (13) using a forward-time and center-space finite differencing to obtain:

$$q(t + \Delta t, A) = \left[1 - \omega\left(L_e^{2+} + L_e^{2-}\right)\right] q(t, A)$$

$$+ \omega L_e^{2+} q(t, A + \Delta A) + \omega L_e^{2-} q(t, A - \Delta A), \quad (15)$$

where

$$L_e^{2\pm} = L_e^2(t, A \pm \Delta A/2), \quad (16)$$

$$\omega \equiv \frac{D\Delta t}{\Delta A^2}. \quad (17)$$

Equation (15) describes the discrete Markov process corresponding to eqn (13). It shows that the likelihood of a tracer molecule finding itself at a point ΔA away over a time interval Δt is ωL_e^2. Thus, given D, a larger L_e^2 provides more opportunity for the tracer molecules to disperse across air mass, whereas small L_e^2 inhibits this from happening. Therefore, L_e^2 is a quantitative measure of *permeability* (leakiness) of the tracer molecules, and one can identify a mixing *barrier* with a local minimum of L_e^2.

To illustrate this, we inject a tracer molecule initially at the center of the domain, and seek the probability distribution for its subsequent position (Fig.5). This is done, essentially, by solving eqn (13) numerically with an implicit method. The left column shows the case with two minima in L_e^2 (one percent of the background value), whereas L_e^2 is uniform in the right column. The molecule finds it difficult to cross the L_e^2 minima and its probability distribution remains more or less confined, and becomes homogenized, between the

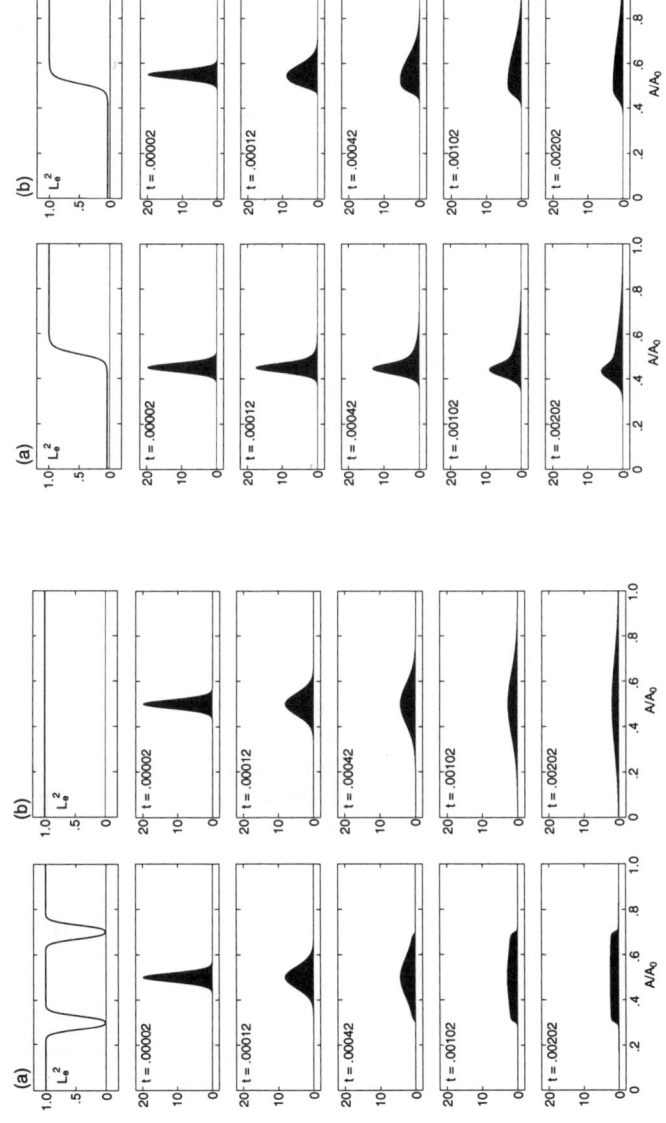

Figure 6 Same as Fig.5 but for a different equivalent length profile and initial placement of the molecule. The top squared equivalent length changes abruptly near the center of the domain. (a) The molecule is initially placed at $A/A_0 = 0.45$. (b) The molecule is initially placed at $A/A_0 = 0.55$.

Figure 5 Evolution of probability distribution of a tracer molecule as a function of the normalized area. The top frames show the profile of the squared equivalent length. The molecule is initially placed at the center. (a) Equivalent length has two minima. (b) Equivalent length is uniform.

two minima. Thus, the region sandwiched by the two minima is a well-defined containment vessel. In the absence of a minimum, however, the probability distribution spreads without bound, hence the definition of a container has to be extended to the whole domain.

When L_e^2 has gradients in A, the likelihood of molecular dispersion becomes asymmetric: the molecules are more likely to move toward the region of larger L_e^2. Thus, if there is a sharp cutoff of L_e^2 in the domain, it can act as a *one-way* barrier to the transport. This is illustrated in Fig.6. There L_e^2 changes abruptly at the center, and a molecule is initially placed at two different locations relative to this "cliff". When placed slightly to the left of the cliff (where L_e^2 is small), the molecule is very likely to escape through the cliff into the region of large L_e^2 (Fig.6a). The molecule placed slightly to the right of the cliff where L_e^2 is large, however, stands less chance of escaping through the cliff, while it is more likely to be dispersed within the region of large L_e^2 (Fig.6b). Hence the equivalent length cliff regulates the direction of mixing for the molecules *in its vicinity*. On the other hand, the molecules placed well interior of the region of small L_e^2 will find it difficult to travel to the cliff. Thus, most of the region of small L_e^2 is well isolated from the region of large L_e^2.

The above examples illustrate that the molecular kinetics of the tracer in the area coordinate can be computed very economically (as a 1D random walk) once equivalent length is known. This is in contrast with the conventional particle tracer methods which are at least in two dimensions and require knowledge of winds. Furthermore, the present method explicitly deal with the kinetics of the tracer molecules, not of the air parcels, which is more pertinent to photochemistry. In the context of the stratospheric transport, the equivalent length diagnostic is particularly useful for quantifying the directionality of planetary wave breaking. Whether the substance is being peeled away from the polar vortex or entrained into it has been discussed in various studies but more or less in a qualitative fashion (Dritschel 1988, Dahlberg and Bowman 1994, M. Nakamura and Plumb 1994, Waugh et al. 1994, Plumb et al. 1994). In the present formalism, this can be diagnosed simply by the sign of the gradient of L_e^2 in the edge regions.

3.4 Edge formation

As noted in section 2, the isentropic distribution of stratospheric tracers in the winter hemisphere is highly inhomogeneous . Various studies have shown that the extreme inhomogeneity, such as the edges, is caused primarily by horizontal mixing, rather than differential vertical advection (Polvani et al. 1995,

Eluszkiewicz *et al.* 1995). Thus eqn (1) should encompass all essential kinematical ingredients for the edge formation, and so does eqn (13), a direct transformation of eqn (1).

We have postulated that the primary effect of chaotic advection is to stretch equivalent length to the point where diffusion effect becomes significant. In the nearly inviscid limit, equivalent length should reflect, therefore, the stretching history of the tracer contour. This prompts us to use eqn (13) as a *kinematic model* of stirring-mixing problem driven by a *prescribed* equivalent length. The spatial and temporal behaviors of equivalent length reflect the nature of stirring. Here we consider a tracer field in a rectangular domain whose area and length are A_0 and L_0, respectively. The initial tracer field is assumed to be

$$q(A,0) \;=\; \cos(\pi A/A_0), \qquad 0 \le A \le A_0. \tag{18}$$

At $t = 0$ stirring is "switched on." The stirring process is represented by an equivalent length that grows exponentially from an initially uniform value (=

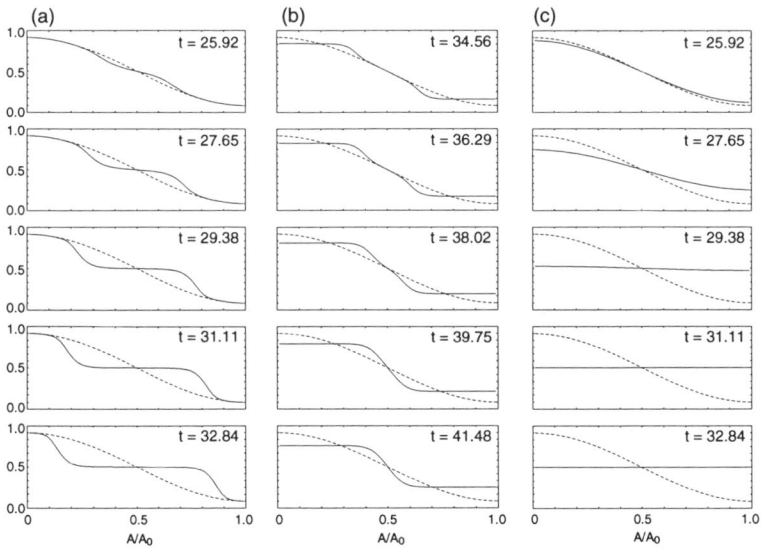

Figure 7 Evolution of the tracer distribution in the 1D kinematic model, plotted as a function of the normalized area. Time is normalized by $\max(s)^{-1}$. Dashed curves indicate the initial condition. (a) Stirring is maximal at the center. (b) Stirring is minimal at the center. (c) Stirring is uniform. See text for details. Adapted from Nakamura (1996). Copyright: American Meteorological Society.

L_0) at rates that vary with A:

$$L_e^2(A,t) = L_0^2 \exp(\gamma(A)t). \tag{19}$$

A nondimensional constant that characterizes this system is

$$\frac{DL_0^2}{\max(\gamma) A_0^2}, \tag{20}$$

which is fixed at 2.0×10^{-13}. Equation (13) is then integrated with time using an implicit scheme, with no-flux boundary conditions at $A = 0$, A_0. In Fig.7 we show the subsequent evolution of the tracer field for different profiles of the stretching rate γ.

The first sequence (left column) is the case of a centralized stirring wherein γ is maximal at the center of the domain and tails off to a smaller background value center where L_e^2 is maximal at the flanks. The shape of γ is gaussian (plus a constant) with characteristic width being 1/3 of the domain shown. The tracer field is essentially unchanged until L_e^2 is significantly stretched. Then, the gradient starts to diminish from the center where L_e^2 is maximal. From that point on, homogenization proceeds quickly and the region of weak gradients gradually spreads sideways. Concomitantly, two distinctive edges are formed at the flanking fronts of the mixed zone. Notice the horizontal scale of the edges is much smaller than that of forcing (gaussian width of γ).

The second sequence concerns a case in which γ is positive but minimal at the center. Thus, stirring is increasingly vigorous away from the center. In this case, homogenization starts from the sides and the gradients are concentrated toward the center.

In both of these examples, the edges do not last very long—without other forcing mechanisms to maintain the gradients, the tracer field will eventually be homogenized globally. Nevertheless, the examples clearly demonstrate that the edges form as a result of differential stirring, that is, spatially varying stretching. This is confirmed by the third sequence in Fig.7 wherein the stretching rate γ is kept uniform. In that case the gradients are destroyed globally from the outset, without an edge being formed. The tracer edges observed in the stratosphere are believed to emerge, at least in part, by the same mechanism. Rapid eddy turnover in the Northern Hemisphere surf zone promotes the sideways erosion of the homogenized air, which squeezes the tracer gradients towards polar and subtropical flanks, much the same way as in the first example above. Meanwhile, Chen (1994) suggests that the kinematics of the Antarctic vortex edge is more like the second example.

It is important to notice that an edge by itself is not necessarily a barrier. While the stationary edge in the second example emerges at the point of minimal L_e^2 so in this case it is indeed a barrier (albeit still leaky), the migrating edges in the first example are more permeable because L_e^2 there is not particularly small. Indeed, the very fact that the edges are migrating implies that mass is being entrained into the surf zone through the edges. This serves as a caveat against the common practice of using the terms *edge* (concentrated gradients) and *barrier* interchangeably. We will discuss more about this point in section 4.3.

3.5 An example of diagnostic

We have so far focused on the conceptual issues of mixing using, at most, a kinematic model. Here we shall apply this diagnostic formalism to a direct numerical simulation of two-dimensional tracer mixing, demonstrating that the foregoing concepts do indeed characterize the essence of mixing.

The example is taken from Nakamura (1996) who studies mixing of heat during the nonlinear life cycle of Kelvin-Helmholtz (KH) instability. [For the mechanics of KH instability see Drazin and Reid (1981) and Scinocca (1995).] Admittedly, KH instability is quite foreign to the planetary-scale mixing in the stratosphere: its horizontal scale is of the order of kilometers or less, and the associated mixing is not isentropic but cross-isentropic. However, the flow kinematics around the KH cat's eye is very similar to that of the planetary-wave nonlinear critical layer in the stratospheric surf zone (Killworth and McIntyre 1985; Haynes 1985). Since the point here is to delineate basic kinematics and not the underlying dynamics, this serves our illustrative purpose and actually attests to the wide applicability of the diagnostic.

Because the design of the simulation is explained in Nakamura (1996), we will not repeat it here. A small-amplitude, normal mode instability of the stratified shear flow is allowed to grow into finite amplitude. In that process, potential temperature is mixed according to eqn (1) except that the streamfunction is defined in the vertical plane. Figure 8 depicts the evolution of potential temperature in two dimensions (left column), also in the area coordinate (middle), and the associated equivalent length of the potential temperature contours (in logarithmic scale, right).

The area analysis of potential temperature and equivalent length is performed as follows. For a given snapshot of potential temperature in the vertical plane, we sample 100 representative contours. We then measure the area that resides between each contour and the lower boundary. From this we construct the area-potential temperature relation. We also compute square of gradient of potential temperature at each grid point and integrate it over the same area that

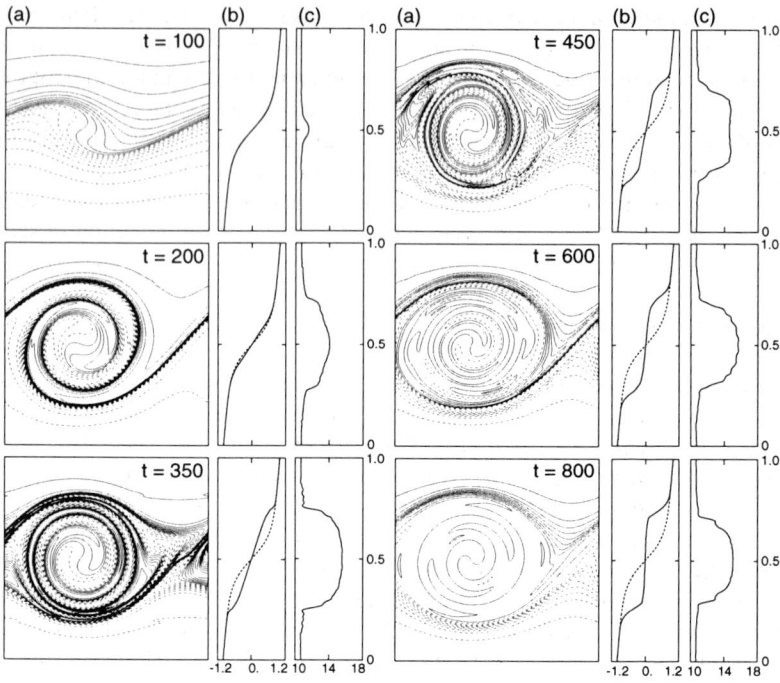

Figure 8 Life cycle of 2D Kelvin-Helmholtz instability in six instants (Unit of time is seconds.) (a) Potential temperature in x and z (interval = 0.1K). The vertical scale is twice that of the horizontal. Dashed curves indicate values less than the median. (b) Potential temperature (abscissa) as a function of area measured relative to the lower boundary and normalized by the area of the domain (ordinate). The dashed curves indicate the initial condition. (c) $\ln L_e^2$ (abscissa) as a function of the normalized area. Adapted from Nakamura (1996).

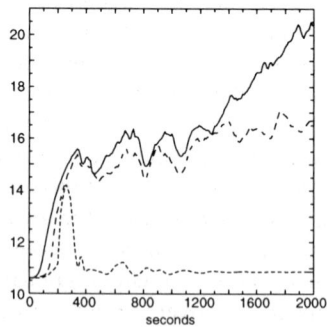

Figure 9 Time dependence of $\ln L_e^2$ during the Kelvin-Helmholtz life cycle shown in Fig.8 at three representative values of normalized area (a). Solid curve: a=0.5. Medium-dashed curve: a=0.625. Fine-dashed curve: a=0.75. The first two lie inside the cat's eye. Adapted from Nakamura (1996).

each potential temperature contour defines. Finally we evaluate eqn (14) numerically.

The contours of potential temperature wrap up rapidly in the middle of the domain as the flow forms a cat's eye. The contours fill the eye region and then get dissipated by diffusion. Eventually potential temperature is homogenized inside the eye, but the concentrated gradients appear along the separatrices of the flow. In contrast, the contours close to the horizontal boundaries remain more or less unperturbed during the course of simulation. Consistent with this, the equivalent lengths of the potential temperature contours increase rapidly in the middle of the domain (Fig.8c). Not only does the magnitude grow, but the region of the growth expands as the cat's eye matures. However, the equivalent lengths outside the eye remain close to the initial value. Hence there emerge two sharp cliffs of equivalent length at the boundaries of the cat's eye (separatrices). As the equivalent lengths grow inside the eye, potential temperature in the area coordinate is homogenized (Fig.8b). The mixing process is very similar to the first sequence of Fig.7 except that the produced edges remain more or less stationary.

Time dependence of equivalent length is depicted in Fig.9 for three representative locations (in terms of area A). The two curves sampled inside the cat's eye exhibit irreversible growth, whereas stretching of the contour outside the eye is at best reversible and hence inconsequential in the long run. Since equivalent length is the ratio between the mean horizontal gradient and areal gradient of the tracer [eqn (14)], it can increase even when the tracer gradients are decaying, as long as the areal gradient decays faster than the mean horizontal gradient. This is why the equivalent length inside the cat's eye does not decay even after most gradients are gone.

3.6 Effect of data resolution

Since the above diagnostic uses the geometrical attributes of tracer contours (area and length) to quantify concepts like *mixedness*, *stirredness* and *barrier*, it is practically constrained by the horizontal resolution of the tracer data. This is particularly true in the case of equivalent length which is a higher order moment of the tracer contours. When the tracer field is well stirred, so much power of equivalent length resides in small scales that, unless the observation resolves these scales, the equivalent length will be severely underestimated. This is not a problem for numerical simulations for which "perfect observation" is possible down to the numerically defined microscale (usually the grid size of the model). Dealing with the real observation, however, one has no choice but to measure "coarse-grained" equivalent length. This brings about an ambiguity in L_e^2 that goes hand in hand with the ambiguity in D. Obviously it is senseless to match the coarse-grained L_e^2 with the molecular diffusivity.

If one assumes that the measured area-tracer relation is unaffected by the data resolution, then eqn (13) can be rewritten as

$$\frac{\partial}{\partial t}q(A,t) = \frac{\partial}{\partial A}\left(D'(A,t;\lambda)\, L_e'^2(A,t;\lambda)\, \frac{\partial q}{\partial A} \right), \tag{21}$$

$$D'(A,t;\lambda)\, L_e'^2(A,t;\lambda) = DL_e^2(A,t), \tag{22}$$

where the prime denotes coarse-grained measurement at resolution λ . To the extent D' is unknown, usefulness of $L_e'^2$ as a diagnostic of mixing is mitigated. Nevertheless, if the *structure*, if not *magnitude*, of $L_e'^2$ retains that of true L_e^2, then the coarse-grain equivalent length can still be used, at least, to identify barriers. One might hope that the contours have self-similar shapes so that a fractal dimension can be utilized to extrapolate the measured length onto the unresolved scales. While in the inertial subrange of homogeneous turbulence this is indeed viable (Batchelor's k^{-1} spectrum, for example, leads to $L_e'^2 \sim \lambda^{-2}$; Batchelor 1959, Nakamura and Ma 1997), in general there is no guarantee that self-similarity prevails universally (Pierrehumbert 1994). It is perhaps more fruitful to use eqn (21) to *diagnose* $D'L_e'^2$ from the measured tracer tendency and areal gradients.

Use of the area-equivalent length diagnostic will be further exploited in the next section where we discuss the vertical cross section of the stratospheric transport.

4 Incorporating diabatic effects

As stated at the beginning of the previous section, the adiabatic vertical transport in the stratosphere is inefficient because of the strong stratification. Thus, a systematic vertical excursion of tracers must be associated with diabatic heating and cooling. Although transport by the diabatically driven circulation is typically slower than the isentropic mixing associated with planetary wave breaking, it cannot be neglected on the seasonal time scale or longer (Hitchman and Leovy 1986, Gille *et al.* 1987).

The diabatic transport is described most concisely if one uses potential temperature θ as the vertical coordinate. Then the heating rate $\dot{\theta}$ becomes the vertical "velocity" that brings about the cross isentropic transport. The summer stratosphere is in a close proximity to radiative equilibrium, so the heating-cooling rates are small. In the winter stratosphere (particularly in the Northern Hemisphere), however, due to the equator-to-pole heat transport, the temperatures in the polar region are kept considerably higher than the radiative equilibrium values. This drives radiative cooling, and hence significant downward tracer transport in the polar region. In the tropics there is compensating upwelling, albeit weak because of the larger area of the region. The resultant hemispheric poleward overturning is known as the Brewer-Dobson circulation (Brewer 1949; Dobson 1956), and tends to steepen the tracer slopes in the meridional plane (Holton 1986; Mahlman *et al.* 1986). While the diabatic effects are traditionally defined in terms of a zonally averaged residual circulation (Dunkerton 1978), here we recast them into the MLM formalism.

4.1 Transport equation in the isentropic mass coordinate

In this section, we generalize the MLM diagnostic to include the effects of diabatic heating, following Nakamura (1995). The model transport equation for the tracer q is, utilizing the isentropic coordinate,

$$\frac{\partial}{\partial t}(\sigma q) = -\nabla_\theta \cdot (\sigma q \mathbf{v}) - \frac{\partial}{\partial \theta}(\sigma q \dot\theta) + \sigma \dot q. \tag{23}$$

In eqn (23) the horizontal divergence and wind vector are defined on the isentropic surface, whereas the density

$$\sigma = -g^{-1}\frac{\partial p}{\partial \theta} \tag{24}$$

defines the thickness of isentropic layer (hydrostatic balance assumed). Here $\dot q$ represents *all* nonconservative sources and sinks for q (i.e., diffusion, photochemical sources, precipitation, etc.), and $\dot\theta$ includes *all* diabatic heating (radiative or otherwise). Mass continuity of the isentropic layer reads

$$\frac{\partial}{\partial t}\sigma = -\nabla_\theta \cdot (\sigma \mathbf{v}) - \frac{\partial}{\partial \theta}(\sigma \dot\theta). \tag{25}$$

We start our inquiry much the same way as in section 2, by associating the tracer isosurfaces and *mass* that resides between the two adjacent isentropic surfaces:

$$\textit{mass in the region } q^* \leq q, \quad \theta \leq \theta^* \leq \theta + \delta\theta \tag{26}$$

The mass element $m(q,\theta,t)$ plays the role of area A in the previous section. Like before, there is one-to-one relationship between m and q on a given isentropic surface, regardless of the geometry of the isosurfaces of q. Therefore, by virtue of an inverse mapping, one can write $q = q(m,\theta,t)$, reducing the spatial dimension to two. We then define the density-weighted area integral of a scalar

$$\mathcal{H}(\,\cdot\,) \equiv \iint_{q^* \le q,\ \theta^* = \theta} (\,\cdot\,)\,\sigma\,dA \;=\; \int_{q_{min}}^{q} dq^* \oint_{q^*} \frac{\sigma}{|\nabla q|}(\,\cdot\,)\,dl. \tag{27}$$

Hence

$$m(q,\theta,t) \;=\; \mathcal{H}(1) \tag{28}$$

and the average of a scalar around the q-contour on the isentropic surface is defined, analogous to eqn (7), as

$$(\tilde{\,\cdot\,}) \;\equiv\; \frac{\partial}{\partial m}\mathcal{H}(\,\cdot\,). \tag{29}$$

The mass continuity in the (q,θ)-space is derived as follows.

$$\frac{\partial m}{\partial t} \;=\; \frac{\partial}{\partial t}\iint_{q^* \le q,\ \theta^* = \theta} \sigma\,dA$$

$$=\; -\frac{\partial}{\partial q}\iint_{q^* \le q,\ \theta^* = \theta}\left(\frac{\partial q}{\partial t}\right)\sigma\,dA \;+\; \iint_{q^* \le q,\ \theta^* = \theta}\frac{\partial\sigma}{\partial t}\,dA \qquad \leftarrow Eqn\ (8)$$

$$=\; -\frac{\partial}{\partial q}\iint_{q^* \le q,\ \theta^* = \theta}\frac{\partial(\sigma q)}{\partial t}\,dA \;+\; q\frac{\partial}{\partial q}\iint_{q^* \le q,\ \theta^* = \theta}\frac{\partial\sigma}{\partial t}\,dA \;+\; \iint_{q^* \le q,\ \theta^* = \theta}\frac{\partial\sigma}{\partial t}\,dA$$

$$=\; \frac{\partial}{\partial q}\left\{ q\iint_{q^* \le q,\ \theta^* = \theta}\frac{\partial\sigma}{\partial t}\,dA \;-\; \iint_{q^* \le q,\ \theta^* = \theta}\frac{\partial(\sigma q)}{\partial t}\,dA \right\}. \tag{30}$$

To derive the third line, we used an identity

$$\frac{\partial}{\partial q} \iint_{q^* \leq q} (\,\cdot\,) q^* \, dA \;=\; \oint_{q^* = q} \frac{(\,\cdot\,) q^*}{|\nabla q|} \, dl \qquad \leftarrow Eqn\ (6)$$

$$=\; q \oint_{q^* = q} \frac{(\,\cdot\,)}{|\nabla q|} \, dl \;=\; q \frac{\partial}{\partial q} \iint_{q^* \leq q} (\,\cdot\,) \, dA \qquad \leftarrow Eqn\ (6). \qquad (31)$$

Substituting eqns (23) and (25) in the last expression of eqn (30), and noting

$$\iint_{q^* \leq q,\, \theta^* = \theta} \nabla_\theta \cdot (\sigma q^* \mathbf{v}) \, dA \;=\; \oint_{q^* = q,\, \theta^* = \theta} \sigma q^* \mathbf{v} \cdot \mathbf{n} \, dl$$

$$=\; q \oint_{q^* = q,\, \theta^* = \theta} \sigma \mathbf{v} \cdot \mathbf{n} \, dl \;=\; q \iint_{q^* \leq q,\, \theta^* = \theta} \nabla_\theta \cdot (\sigma \mathbf{v}) \, dA, \qquad (32)$$

$$\frac{\partial}{\partial \theta} \iint_{q^* \leq q,\, \theta^* = \theta} \sigma \dot\theta \, dA \;=\; \iint_{q^* \leq q,\, \theta^* = \theta} \frac{\partial(\sigma \dot\theta)}{\partial \theta} \, dA$$

$$-\; \frac{\partial}{\partial q} \iint_{q^* \leq q,\, \theta^* = \theta} \sigma \dot\theta \frac{\partial q}{\partial \theta} \, dA, \qquad \leftarrow Eqn\ (8) \quad (33)$$

etc., eqn (30) finally reduces to

$$\left. \frac{\partial m}{\partial t} \right|_{q,\theta} \;=\; -\frac{\partial}{\partial q} \iint_{q^* \leq q,\, \theta^* = \theta} \sigma \dot q \, dA \;-\; \frac{\partial}{\partial \theta} \iint_{q^* \leq q,\, \theta^* = \theta} \sigma \dot\theta \, dA$$

$$=\; -\left. \frac{\partial \mathscr{H}(\dot q)}{\partial q} \right|_{\theta,t} \;-\; \left. \frac{\partial \mathscr{H}(\dot\theta)}{\partial \theta} \right|_{q,t}, \qquad (34)$$

where the subscripts denote the parameters to be fixed during the partial derivative operations. Equation (34) describes the mass budget of a "shell" en

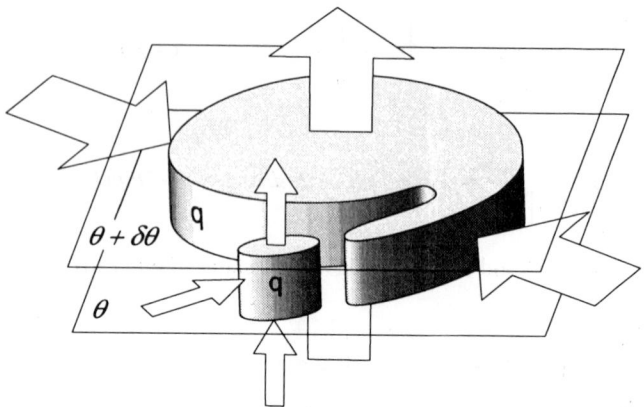

Figure 10 Schematic view of the mass budget of a "shell" enclosed by an
isosurface of quasi-conservative tracer q between adjacent isentropic surfaces.
"Islands" are counted as part of the main shell. Arrows indicate mass fluxes through
the bounding surfaces due to nonconservative processes. The mass of the shell is
$m\delta\theta$. After Nakamura (1995). Copyright: American Meteorological Society.

closed by two adjacent isentropic surfaces and isosurface of the tracer (Fig.10).
The first term on the right hand side of eqn (34) represents the sideways mass
flux convergence through the tracer isosurface. The second term represents the
convergence of the diabatic mass flux through the isentropic surfaces. These
terms vanish if the flow is adiabatic and the tracer is conserved following the
motion of air. Notice if eqn (34) is integrated with respect to θ over the closed
isosurface of tracer q, the contribution from the diabatic term vanishes. This
means the *net* gain or loss of mass of the container (e.g. shrinkage of the polar
vortex) is solely due to the tracer nonconservation. Now the equation for the
tracer transport relative to m is obtained by inverting eqn (34):

$$\left.\frac{\partial q}{\partial t}\right|_{m,\theta} = -\left.\frac{\partial m}{\partial t}\right|_{q,\theta} \times \left.\frac{\partial q}{\partial m}\right|_{\theta,t}$$

$$= \left(\left.\frac{\partial \mathcal{H}(\dot{q})}{\partial q}\right|_{\theta,t} + \left.\frac{\partial \mathcal{H}(\dot{\theta})}{\partial \theta}\right|_{q,t}\right) \times \left.\frac{\partial q}{\partial m}\right|_{\theta,t}$$

$$
= \left(\left. \frac{\partial \mathscr{M}(\dot{q})}{\partial q} \right|_{\theta,t} + \left. \frac{\partial \mathscr{M}(\dot{\theta})}{\partial \theta} \right|_{m,t} + \left. \frac{\partial \mathscr{M}(\dot{\theta})}{\partial m} \right|_{\theta,t} \times \left. \frac{\partial m}{\partial \theta} \right|_{q,t} \right) \times \left. \frac{\partial q}{\partial m} \right|_{\theta,t}
$$

$$
= \left(\left. \frac{\partial \mathscr{M}(\dot{q})}{\partial q} \right|_{\theta,t} + \left. \frac{\partial \mathscr{M}(\dot{\theta})}{\partial \theta} \right|_{m,t} \right) \times \left. \frac{\partial q}{\partial m} \right|_{\theta,t} - \left. \frac{\partial \mathscr{M}(\dot{\theta})}{\partial m} \right|_{\theta,t} \times \left. \frac{\partial q}{\partial \theta} \right|_{m,t} . \qquad (35)
$$

We used eqn (34) in the second line above. Equation (35) can be further rewritten:

$$
\frac{Dq}{Dt} = \left. \frac{\partial q}{\partial t} \right|_{m,\theta} + \left. \frac{Dm}{Dt} \frac{\partial q}{\partial m} \right|_{\theta,t} + \left. \frac{D\theta}{Dt} \frac{\partial q}{\partial \theta} \right|_{m,t} = 0, \qquad (36)
$$

$$
\frac{Dm}{Dt} = - \left. \frac{\partial \mathscr{M}(\dot{q})}{\partial q} \right|_{\theta,t} - \left. \frac{\partial \mathscr{M}(\dot{\theta})}{\partial \theta} \right|_{m,t}, \qquad (37)
$$

$$
\frac{D\theta}{Dt} = \left. \frac{\partial \mathscr{M}(\dot{\theta})}{\partial m} \right|_{\theta,t} = \tilde{\dot{\theta}}. \qquad (38)
$$

Here the material derivative D/Dt is defined following the two-dimensional motion of the tracer contour in the (m,θ)-space. The effective transport velocity ($Dm/Dt, D\theta/Dt$) is the nonconservative mass flow that "advects" the tracer contours. This includes both the mass flow due to the diabatic heating (Brewer-Dobson circulation) represented by the $\dot{\theta}$ terms and that due to the tracer nonconservation ($\dot{q}$ term). Notice the diabatic mass flow is nondivergent. This is consistent with the conservation of mass enclosed by the tracer isosurface as noted previously. Divergence of the mass flow is therefore solely associated with the tracer nonconservation. In other words, the effect of diabatic circulation is to redistribute mass *along* the tracer isosurfaces, whereas that of tracer nonconservation is to redistribute mass *across* them. This suggests that a formation of extreme and cohesive tracer inhomogeneity in the (m,θ)-space is primarily driven by the divergent part of the mass flow, namely the tracer nonconservation, rather than by the diabatic effect.

Now we rewrite the $\dot{q}$ term as a sum of isentropic diffusion and everything else, that is,

$$\dot{q} \;=\; \frac{D}{\sigma} \nabla_\theta \cdot \left(\sigma \nabla_\theta q \right) \;+\; S. \tag{39}$$

Then, a manipulation similar to what led to eqn (13) transforms eqn (36) to

$$\frac{\partial}{\partial t} q(m, \theta, t) \;+\; J_{m\theta}\!\left(\mathcal{H}(\dot{\theta}), q \right) \;=\; D \frac{\partial}{\partial m}\!\left(K^2 \frac{\partial q}{\partial m} \right) \;+\; \tilde{S}, \tag{40}$$

$$J_{m\theta}(x, y) \;=\; -\frac{\partial x}{\partial \theta}\frac{\partial y}{\partial m} + \frac{\partial x}{\partial m}\frac{\partial y}{\partial \theta}, \tag{41}$$

$$K^2(m, \theta, t) \;=\; \left| \nabla_\theta \tilde{q} \right|^2 \left(\partial q / \partial m \right)^{-2}. \tag{42}$$

Here K is the *density-weighted* equivalent length which, in effect, is the area of the bounding tracer isosurface of the shell in Fig.10. The second term on the left hand side of eqn (40) represents advection by the diabatic mass flow, while the right hand side represents isentropic mixing and other nonconservative processes.

4.2 Scaling of mass

Equation (40) is a natural extension of eqn (13) in two dimensions. Unfortunately, mass element m is not very useful for covering the entire stratosphere: Since the density σ decreases exponentially with height, m in the upper stratosphere is orders of magnitude smaller than in the lower stratosphere (for the same q). This means the measure of "latitude" depends on height—an undesirable feature.

To alleviate this problem, we introduce *equivalent area* A_e and *equivalent latitude* φ_e as

$$A_e(m, \theta, t) \;=\; m / \rho_\theta \;=\; 2\pi a^2 (1 - \sin\varphi_e), \tag{43}$$

$$\rho_\theta \;\equiv\; \sigma_0 \left(\theta / \theta_0 \right)^{-\alpha}, \tag{44}$$

where a is the earth's radius, while σ_0 and θ_0 are mean density and potential temperature, respectively, at some reference level. The power of the normalized potential temperature in eqn (44) is meant to offset the exponential decay of m with height. The equivalent area A_e is then the area that m would cover if its density were everywhere uniform on the isentropic surface and equal to ρ_θ.

It is different from the actual area enclosed by the tracer contour, but the difference can be minimized globally if α is chosen judiciously. For an isothermal atmosphere α is exactly 4.5 since density scales as $\sigma \sim \theta^{-4.5}$ (Lait 1994). However, considering the increasing tendency of temperature in the lower and middle stratosphere, a somewhat smaller value of α would be more representative. We will use $\alpha = 4.0$ in the GCM analysis described later.

The equivalent latitude φ_e is the bounding latitude of a circle centered at the pole whose enclosed area is A_e. Physically, this is the latitude at which a tracer contour would be found if it had no waviness and if the density were uniform on the isentropic surface. The coordinate transformation (43) relates mass element m and equivalent latitude φ_e one to one. Using φ_e instead of m as the meridional coordinate, eqn (40) can be rewritten as

$$\frac{\partial}{\partial t} q(\varphi_e, \theta, t) + \frac{\tilde{v}}{a} \frac{\partial q}{\partial \varphi_e} + \dot{\tilde{\theta}} \frac{\partial q}{\partial \theta}$$

$$= \frac{D}{4 \pi^2 a^4 \rho_\theta^2 \cos \varphi_e} \frac{\partial}{\partial \varphi_e} \left(\frac{K^2}{\cos \varphi_e} \frac{\partial q}{\partial \varphi_e} \right) + \tilde{S}, \qquad (45)$$

$$\tilde{v} \equiv \frac{1}{2 \pi a \rho_\theta \cos \varphi_e} \left. \frac{\partial \mathcal{M}(\dot{\theta})}{\partial \theta} \right|_{\varphi_e, t}, \qquad (46)$$

$$\dot{\tilde{\theta}} \equiv - \frac{1}{2 \pi a^2 \rho_\theta \cos \varphi_e} \left. \frac{\partial \mathcal{M}(\dot{\theta})}{\partial \varphi_e} \right|_{\theta, t}. \qquad (47)$$

Nondivergence of the diabatic mass flow is unaffected by this transformation:

$$\frac{1}{a \cos \varphi_e} \frac{\partial}{\partial \varphi_e} (\tilde{v} \cos \varphi_e) + \frac{1}{\rho_\theta} \frac{\partial}{\partial \theta} \left(\rho_\theta \dot{\tilde{\theta}} \right) = 0. \qquad (48)$$

The effect of the diabatic circulation in eqn (45) is to carry the tracer molecules with mass of air, while that of horizontal mixing on the right hand side is to horizontally disperse tracer molecules across the air mass.

It is useful to note the link between the above formulation and the area diagnostics of Butchart and Remsberg (1986). In the stratosphere, the isentropic winds are nearly nondivergent and the density of air depends, to the lowest order, only on potential temperature:

$$\sigma \;\approx\; \sigma(\theta). \tag{49}$$

Under this approximation, from eqns (4) and (28),

$$\mathcal{M}(\,\cdot\,) \;\approx\; \sigma(\theta)\,\mathcal{A}(\,\cdot\,), \tag{50}$$

$$m(q,\theta,t) \;\approx\; \sigma(\theta)\,A(q,\theta,t). \tag{51}$$

In this case equivalent latitude can be redefined in terms of the *actual* area enclosed by the tracer contours:

$$A(q,\theta,t) \;=\; m/\sigma \;=\; 2\pi a^2\!\left(1-\sin\varphi_e\right). \tag{52}$$

Correspondingly, eqns (45)-(48) are approximated as:

$$\frac{\partial}{\partial t}q(\varphi_e,\theta,t) \;+\; \frac{\hat{v}}{a}\frac{\partial q}{\partial \varphi_e} \;+\; \hat{\dot{\theta}}\,\frac{\partial q}{\partial \theta}$$

$$=\; \frac{D}{4\pi^2 a^4 \cos\varphi_e}\,\frac{\partial}{\partial\varphi_e}\!\left(\frac{L_e^2}{\cos\varphi_e}\frac{\partial q}{\partial\varphi_e}\right) \;+\; \hat{S}, \tag{53}$$

$$\hat{v} \;\equiv\; \frac{1}{2\pi\, a\cos\varphi_e\,\sigma}\,\frac{\partial\sigma\,\mathcal{A}(\dot{\theta})}{\partial\theta}\bigg|_{\varphi_e,t}, \tag{54}$$

$$\hat{\dot{\theta}} \;=\; -\frac{1}{2\pi\, a^2\cos\varphi_e}\,\frac{\partial\mathcal{A}(\dot{\theta})}{\partial\varphi_e}\bigg|_{\theta,t}, \tag{55}$$

$$\frac{1}{a\cos\varphi_e}\,\frac{\partial}{\partial\varphi_e}\big(\hat{v}\cos\varphi_e\big) \;+\; \frac{1}{\sigma}\frac{\partial}{\partial\theta}\!\left(\sigma\hat{\dot{\theta}}\right) \;=\; 0. \tag{56}$$

Here L_e is defined by eqn (14). Equations (53)-(56) approximately govern the tracer distribution in the space spanned by the area equivalent latitude and potential temperature. Hence this corresponds to the area analysis of Butchart and Remsberg (1986), though they only considered the diabatic effects on potential vorticity. The above MLM formalism thus represents a generalization to the Butchart-Remsberg area analysis with an explicit representation of horizontal mixing.

4.3 MLM analysis of N_2O in SKYHI GCM

To demonstrate the utility of the above formalism, the N_2O mixing ratio simulated by the GFDL SKYHI GCM is now analyzed. The outputs of a GCM grant "perfect observation" to within the numerical resolution, and variables are internally consistent. Therefore, they are more suitable for the purpose of diagnostic test than satellite data or airborne observations that are limited in accuracy, resolution and coverage. Comparison with the Upper Atmosphere Research Satellite measurements will be discussed elsewhere (Nakamura and Ma 1997). This GCM covers the entire globe up to 80km in altitude with a horizontal resolution of 1.2 degrees (longitude) and 1 degree (latitude) on 40 sigma levels. The design and general climatology of the model are found, for example, in Fels *et al.* (1980), Mahlman and Umscheid (1987), Strahan and Mahlman (1994ab), Hamilton *et al.* (1995), and Hamilton (1995), so they will not be repeated here. The N_2O field was initialized by interpolation from an October condition of the lower-resolution run. Its photodissociation rates in the upper stratosphere are prescribed, but the model neglects the N_2O source near the ground. This results in a weak net loss of N_2O over the domain, but in the middle to lower stratosphere, its chemical life time is much longer than the dynamical time scale.

Usage of eqn (45) or (53) depends on the availability of the observed variables. In the following we only use the N_2O mixing ratio and density σ as observed data, although the model output includes all other terms including $\dot{\theta}$. While this may be seen as under-exploitation of data, it serves as a surrogate satellite observation in which heating rates are not readily available. A series of MLM cross sections of N_2O in the (φ_e, θ)-space will be constructed, along with the equivalent length distribution. Diabatic effect is then *inferred* from the vertical excursion of the tracer contours outside the surf zone.

Processing of the N_2O mixing ratio for the MLM diagnosis is similar to that outlined in section 3.5. Here we include density-weighting in the area integral and repeat the whole process for all isentropic layers. We use $\theta_0 = 811K$, $\sigma_0 = 4.46 \times 10^{-1}\ kg\ m^{-2} K^{-1}$, and $\alpha = 4$ in eqn (44).

4.3.1 Comparison with Eulerian zonal mean

Figure 11 compares the vertical cross section of the N_2O mixing ratio between the conventional zonal mean analysis and the MLM analysis. The figure is a snapshot of the Northern Hemisphere stratosphere during late winter of the model simulation—the same instant as in Fig.2. The three plots share the ordinate which is potential temperature in a logarithmic scale, roughly covering the entire stratosphere. The zonal mean plot uses geographical latitude as the hori-

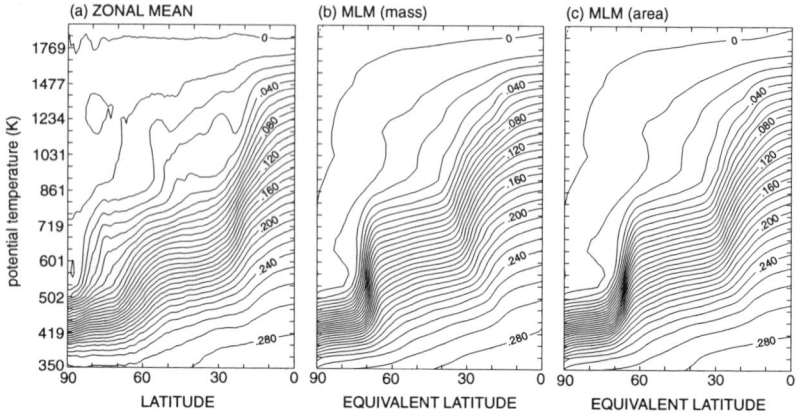

Figure 11 (a) Zonally averaged N_2O mixing ratio in the Northern Hemisphere stratosphere at 1200 UTC 2 February 1984 (model year), simulated by GFDL SKYHI GCM. Potential temperature in the ordinate is in a logarithmic scale. (b) The same instantaneous N_2O field as (a) but based on the MLM analysis. The equivalent latitude in the abscissa is defined by eqn (43). (c) Same as (b) but using the area analysis of Butchart and Remsberg (1986). The equivalent latitude is given by eqn (52). In all figures the contour interval is 0.01ppm. After Nakamura (1995). Copyright: American Meteorological Society.

zontal coordinate, whereas the two MLM plots use equivalent latitude.

The zonal-mean plot (left) captures the general tendency of the N_2O mixing ratio—decreasing toward higher altitudes and latitudes. However, the variation is rather smooth, and the existence of the two edges that are evident in Fig.2 is only hinted, but not fully resolved. This is because the zonal averaging smears the zonally asymmetric edges, as pointed out in section 2.

The two right plots of Fig.11 are the MLM analyses of the same data, using the exact formulation [eqn (43), middle] and approximate method [eqn (52), right]. The approximate method is essentially a layerwise application of the Butchart and Remsberg area analysis. In both plots, the sharp transitions among the polar vortex, surf zone and tropical reservoir are superbly resolved in almost stepwise variation of the tracer slopes. The exact and approximate analyses yield qualitatively similar results, attesting to the validity of the approximation made in eqn (49). The appreciable differences between the two—slightly steeper polar edge and larger tropical reservoir in the exact analysis—may reflect the decreasing tendency of σ with equivalent latitude, but a less than optimal value of α (here chosen to be 4) can also introduce this systematic discrepancy.

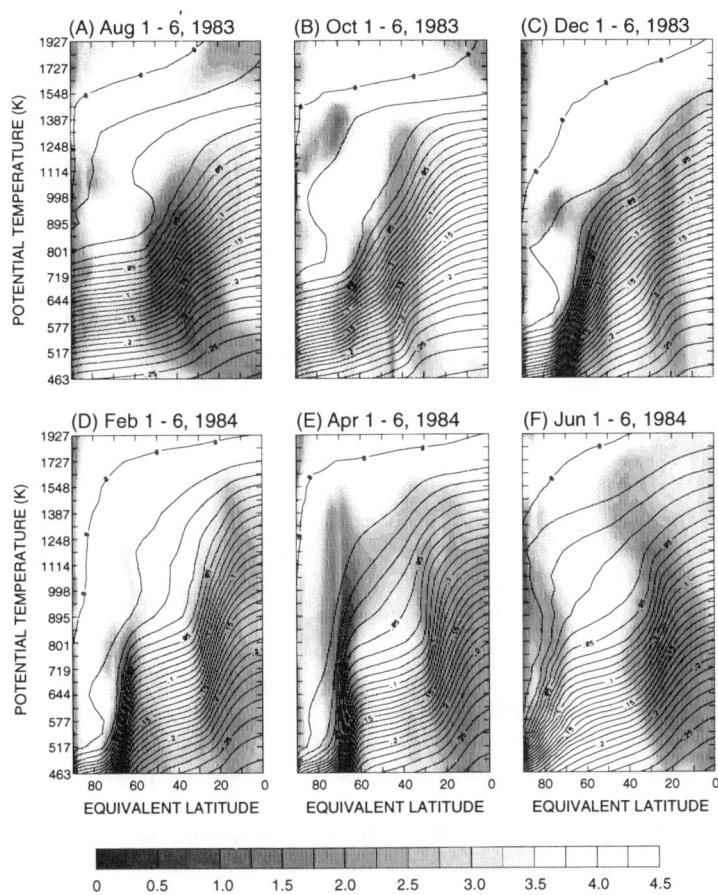

Figure 12 Seasonal variation of the MLM cross section of the N_2O mixing ratio (solid contours, interval = 0.01ppm) and log of normalized equivalent length (gray scale, see text for definition) in the Northern Hemisphere, simulated by GFDL SKYHI GCM. Each frame represents an average of six daily snapshots, plotted against area equivalent latitude (abscissa) and potential temperature (ordinate). (a) August 1 - 6, 1983 (model year). (b) October 1 - 6, 1983. (c) December 1 - 6, 1983. (d) February 1 - 6, 1984. (e) April 1 - 6, 1984. (f) June 1 - 6, 1984.

Typically, the tracer mixing ratio changes more slowly with respect to equivalent latitude than to the Eulerian horizontal coordinates. This is because leakage of mass through the tracer isosurfaces is a much slower process than the fluctuation in the *shape* of the isosurfaces (except during wave breaking events of an unusual magnitude). Therefore, it is unnecessary to take time average to define a robust MLM cross section: even a snapshot like Fig.11 is representative of the state over a duration of period (say, order of a week). Without having to process multiple files, one can greatly economize the data analysis.

4.3.2 Seasonal variability and barrier migration

Figure 12 shows the seasonal variation of the MLM N_2O mixing ratio in the Northern Hemisphere stratosphere. Here the approximate area analysis [eqn (53)] is used instead of the full analysis. Superposed on the MLM cross section is the gray scale representation of the normalized squared equivalent length in logarithmic scale:

$$\ln\left(L_e^{\,2}\big/L_0^{\,2}\right) \;\equiv\; \xi, \tag{57}$$

where L_0 is the length of the zonal circle at the corresponding equivalent latitude. Literally, ξ is a measure of the "stretchedness" of the tracer contour which vanishes if the tracer contour is perfectly zonal. It is also a measure of leakiness of the tracer molecules in the same sense as defined in section 3. A local minimum in ξ (denoted by dark gray) identifies a barrier to the horizontal mixing.

In general, the mixing barriers (minimal ξ) in Fig.12 are found where the lateral tracer gradients are maximal (edges). This may appear to be a simple consequence of the inverse proportionality of L_e and $\left|\partial q/\partial A\right|$ [eqn (14)], but since the other factor $\left|\nabla q\right|$ is independent of $\left|\partial q/\partial A\right|$, the edge-barrier collocation is more than a mathematical triviality. It actually suggests that the fast mixing on both sides of the barrier causes the tracer gradients to concentrate toward the barrier, just as Fig.7b demonstrates.

In summer, there is a broad zone of minimal ξ in the midlatitudes, with a modest equator-to-pole variation in the N_2O mixing ratio (Fig.12a). The tracer contours in the polar region begin to descend rapidly through autumn to winter, due to radiative cooling (Fig.12b,c). Meanwhile, the barrier splits in two meridionally and their separation becomes more apparent towards the late winter (Fig.12d). Concomitantly, the horizontal stirring is enhanced in the midlatitudes (large values of ξ) and the tracer gradients are reduced between the two barriers (formation of the *surf zone*). As a result, the tracer gradients are concentrated in the vicinity of the narrow Arctic barrier (the edge of the *polar vortex*) and the broader subtropical barrier. This is similar to the edge formation process described in section 3.5 (Fig.7a). Here, however, the sharpness of the polar edge is further enhanced by both the fast descending motion and significant stirring inside the polar vortex. The N_2O molecules would be dispersed easily within the surf zone and polar vortex where ξ is relatively large, but not across the barriers. In this sense, the polar vortex, surf zone and tropical reservoir are indeed well-defined containment vessels of the N_2O molecules. The smaller the values of ξ at the barriers, the more isolated are the containers from each other.

Expansion of the surf zone towards the late winter reflects the entrainment of mass from the polar vortex and tropical reservoir by means of repeated planetary wave breaking events. As the surf zone erodes sideways, the polar vortex becomes compact while the barrier at the edge appears to become taller (Fig.12e). The polar barrier and edge altogether begin to disappear at the demise of the vortex in late spring (Fig.12f). Here the horizontal mixing alone destroys much of the equator-to-pole tracer variation, but additional lift of the tracer isosurfaces in the polar region is accomplished by the diabatic circulation which by this time turns upward due to the heating of the ozone layer. The circumpolar circulation changes from westerly to easterly in summer, shutting off the vertical propagation of the planetary waves (Charney and Drazin 1961). Without a further supply of large-scale stirring, rapid mixing in the extratropics soon comes to a halt. The subtropical barrier moves toward the midlatitudes, while the remnant of the surf zone air is effectively "frozen" in the polar region (Fig.12f,a)

Since the life span of the polar barrier is less than a year, the entire extratropics of the stratosphere should be well mixed over the time scale of a year or longer. However, the subtropical barrier is perennial, and thus the N_2O molecules in the tropical air may be well isolated from the extratropics, consistent with the "tropical pipe" concept of Plumb (1996). [In the figure the tropical barrier appears much weaker than the polar barrier (larger ξ). This is partly due to the small scale noise in the tracer contours that is pronounced in the model's tropics (see Fig.2). Much of this noise is believed to be numerical, so the leakiness of the tropical barrier may be overestimated.]

Inside the polar vortex, the summer-to-winter vertical excursion of the tracer contours is about 300K potential temperature levels in the lower stratosphere. This translates to a net cooling rate of about 0.6K/day (in temperature) at 25hPa surface, smaller than the recent observational estimates (e.g. Geller *et al.* 1992, Rosenfield *et al.* 1994, Eluszkiewicz *et al.* 1996) by up to a factor of 2. The cold pole bias of the GCM is suspected as a primary reason for this.

4.3.3 Interannual and interhemispheric variabilities
The above annual cycle of the stratospheric N_2O mixing ratio is, apart from the fact that it is a model simulation, only pertinent to the Northern Hemisphere for

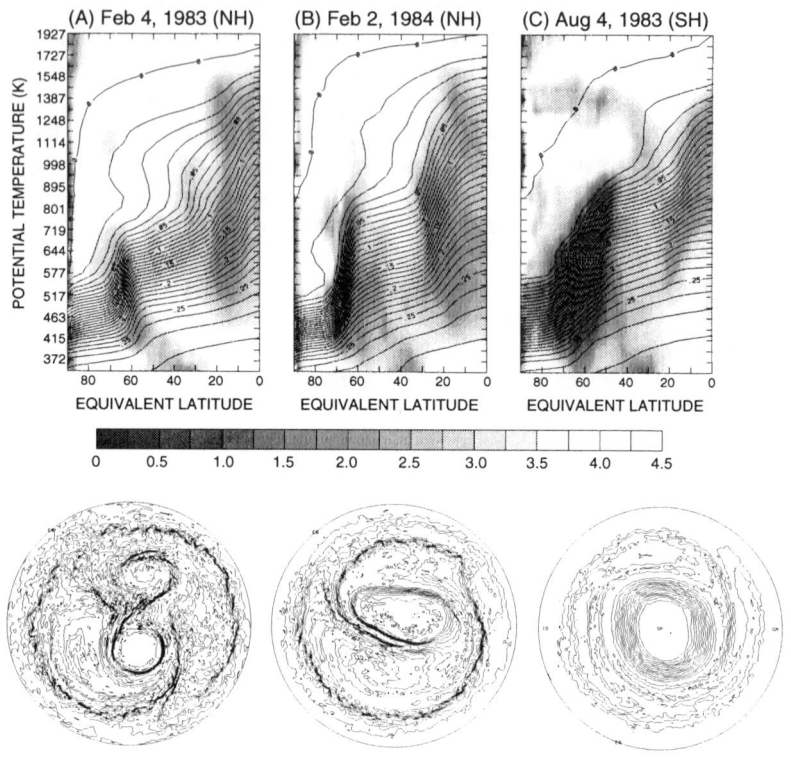

Figure 13 Snapshots of the N_2O mixing ratio for three distinct states of the polar vortex simulated by GFDL SKYHI GCM. (a) February 4, 1983 (model year), Northern Hemisphere. (b) February 2, 1984, Northern Hemisphere. (c) August 4, 1984, Southern Hemisphere. Upper frames: MLM cross sections similar to those of Fig.12. Lower frames: Azimuthal equidistant projections of the N_2O mixing ratio on the 650K isentropic surface corresponding to the upper frames (interval = 0.012ppm). The outer edge of each frame is the equator.

one particular year. Given that the SKYHI simulation at this resolution has been performed to date only for a few model years, we cannot fully address the climatology and variability of the stratospheric transport from this data alone. Nevertheless, hints of natural variability abound even within the first two-year period of simulation.

Among others, interannual and interhemispheric variabilities are the most evident. This is illustrated in Fig.13 where three distinct states of the winter polar vortex are sampled in the N_2O mixing ratio. The first case is the snapshot of N_2O in the Northern Hemisphere amid a major warming event that occurred during the first model winter (see Mahlman and Umscheid 1987 for the documentation of this simulated event). The second is similar to the first except that it is sampled during the second model winter (the same winter as in Fig.12). This winter was found "normal" in the sense that no major warming event was observed. The last is the Southern Hemisphere snapshot for the equivalent season. The MLM cross sections (upper panels) are accompanied by the horizontal projection of the N_2O mixing ratio on the 650K isentropic surface (lower panels) for the respective dates.

Comparison between the major warming and non-warming situations reveals much different vertical and horizontal cross sections. The edge of the polar vortex during the major warming exhibits a peculiar "8" letter shape on the 650K surface, reflecting the splitting of the polar vortex. Similar behavior has been identified by others in potential vorticity during numerically simulated sudden warmings (e.g. Matsuno 1971, Dunkerton et al. 1981, Dunkerton and Delisi 1986). However, this edge appears quite shallow in the MLM cross section in that the variation of the N_2O mixing ratio across the edge is much less than in the non-warming situation, and the edge is altogether absent above 720K. As a result, the height of the N_2O contours during the warming event tend to be higher (lower) in the polar vortex (surf zone) compared with those in the non-warming case. The weaker contrast in the N_2O mixing ratio across the vortex edge during the major warming suggests that the horizontal mixing is very efficient. This is certainly consistent with the more contorted N_2O contours on the isentropic surface and the correspondingly leakier polar barrier (the minimal ξ at the vortex edge is much larger). However, it is not too clear whether these differences in the overall tracer morphology are created by this particular warming event alone or they have been set up by the successive wave breaking events preceding this major warming event. Finally, the width of the surf zone during the warming event is appreciably greater than in the non-warming winter, with a distinct "tongue" of large ξ extending up from the lower tropics. This attests to an enhanced eddy forcing (and associated mixing) applied from the troposphere during the warming event.

The N_2O distribution in the Southern Hemisphere winter is very distinct from that in the Northern Hemisphere. Here the planetary wave forcing is weak, hence the tracer contours in the polar vortex are remarkably zonally symmetric. As a result, ξ is very nearly zero everywhere in the polar vortex. Therefore, the molecules in the Antarctic vortex are highly isolated and unlikely to escape except those very close to the edge (cf. Fig.6). The situation is in contrast with the normal Northern Hemisphere winter in which ξ is significantly larger inside the vortex. Consequently, the tracer slopes inside the Arctic vortex are flatter than in the Antarctic vortex. The mixing property across the Arctic vortex edge is thus analogous to the second sequence of Fig.7. Consistent with the weaker wave forcing, the surf zone in the Southern Hemisphere is underdeveloped. This makes the sideways erosion (section 3.5) less effective in sharpening the polar vortex edge and stripping mass from the polar vortex. Thus, the size and cohesiveness of the Antarctic vortex are appreciably greater than those of the Arctic vortex, while the gradient concentration at the vortex edge is less pronounced.

Another important interannual variability that may exist in the atmosphere is that associated with the phase of Quasi Biennial Oscillation (QBO; Holton and Austin 1991, Baldwin and Tung 1994). The SKYHI GCM does not resolve QBO, thus we cannot address its effect from this dataset. However, the present method should in principle be equally useful for diagnosing the variability of the tropical barrier in response to the QBO phases.

5 Note on potential vorticity and circulation

We have so far concentrated on the kinematics of a passive (chemical) tracer, using diabatic heating rates and equivalent lengths as two external parameters that drive the transport in the MLM cross section. These parameters are in turn regulated by the dynamical state of the containment vessel, namely, its position and surrounding flow. Hence, to gain a deeper understanding of the transport drives, we must ultimately understand the dynamical state of the stratosphere. In this regard it is important to note the role of potential vorticity (PV) as a dynamical tracer (Hoskins $et\ al.$ 1985).

Here we apply the MLM equation (36) to PV and delineate how transport processes affect the mean flow. In the context of the MLM formalism, the mean flow surrounding the containment vessel is naturally represented by $circulation$. Let $C(m,\theta,t)$ be the circulation around the contour of PV that encloses mass element m on an isentropic surface. Here PV is defined as

$$P = \zeta/\sigma, \tag{58}$$

and ζ is the absolute isentropic vorticity (Andrews $et\ al.$ 1987). Then C is the density-weighted area integral of P on the isentropic surface (Hoskins 1991):

$$C(m,\theta,t) = \mathscr{H}(P) = \int_0^m P(m^*,\theta,t)\,dm^*, \tag{59}$$

where $q \equiv P$ is assumed in eqn (27). Substituting $q = P$ in eqn (36) and integrating with respect to m, one obtains

$$\left. \frac{\partial C}{\partial t} \right|_{m,\theta} = \left(\mathcal{H}\left(\dot{P}\right) + \left. \frac{\partial \mathcal{H}\left(\dot{\theta}\right)}{\partial \theta} \right|_{m,\theta} P \right) - \frac{\partial}{\partial \theta} \int\limits_{0}^{m} \tilde{\theta} \, P \, dm^{*}$$

$$\equiv F. \qquad (60)$$

The bracket term on the right hand side of eqn (60) denotes the sideways flux of PV into the shell defined in Fig.10, whereas the second term is the vertical convergence of PV flux across the bounding isentropic surfaces. If the flow is frictionless and adiabatic, $\dot{P} = \dot{\theta} = 0$ hence the right hand side of (60) vanishes. On a rotating sphere, eqn (60) becomes

$$\left. \frac{\partial C_r}{\partial t} \right|_{m,\theta} = -2\Omega \left. \frac{\partial A'}{\partial t} \right|_{m,\theta} + F, \qquad (61)$$

where C_r and Ω are the relative circulation and (constant) rotation rate, respectively, while A' is the area of the PV shell projected onto the equatorial plane. Equation (61) really is the Bjerknes circulation theorem, but, since C_r is the wind integrated around the PV contour, it can also be thought of as the equation for the mean flow or *nonacceleration theorem* (Charney and Drazin 1961, Andrews *et al.* 1987, Ch.3).

The relative circulation can be modified by either change in A' or the nonconservative effect F. The A' term on the right hand side of eqn (61) is formally related to the Coriolis acceleration associated with the mean residual circulation ($f \bar{v}^{*}$) in other formalisms. Here A' changes when the PV shell drifts meridionally or when it is vertically stretched. For example, when the planetary wavenumber one is amplified, the polar vortex is displaced to a lower latitude, decreasing A'. Hence the cyclonic circulation around the edge of the polar vortex *increases*.

Meanwhile, horizontal mixing tends to decelerate circulation through F. To appreciate this, consider a strictly two-dimensional mixing in which vorticity ζ and area A replace P and m, respectively, while $\dot{\theta} \equiv 0$. Assuming that vorticity is diffused as $\dot{\zeta} = D\nabla^{2}\zeta$, eqn (60) reduces to

$$\left.\frac{\partial C}{\partial t}\right|_{A,\theta} = \mathcal{A}\!\left(\dot{\zeta}\right) = D\!\left(\frac{\partial \zeta}{\partial A}\right)^{-1}|\hat{\nabla \zeta}|^{2}, \qquad (62)$$

where eqn (12) was used. For a cyclonic vorticity shell, $\zeta > 0$, $C > 0$, and $\partial \zeta/\partial A < 0$, and thus the right hand side of eqn (62) works as damping to the circulation (same conclusion holds for an anticyclonic shell). This can also be thought of as dissipation of pseudomomentum by diffusion integrated around the vorticity contour.

In comparison with other forms of nonacceleration theorem, characteristically absent from eqn (61) is the "wave-mean flow interaction" term or the Eliassen-Palm flux divergence (Andrews *et al.* 1987). This is because our mean field absorbs what would be *waves* in other formalisms. Since a flavor of wave-mean flow interaction is lost, we cannot use the standard wave dynamics to interpret the *dynamical forcing* to the mean flow and parameterize planetary wave breaking the same way as other mean transport models do (e.g. Yang *et al.* 1990, Garcia 1991). An altogether different approach would be required for parameterizing the two terms on the right hand side of eqn (61).

6 Conclusions

In this article, we have reviewed the development of the modified Lagrangian-mean (MLM) diagnostic formalism of the stratospheric tracer transport. Although no new physics has been introduced beyond what is already known, the mean transport has been recast in a Lagrangian vertical cross section, which grants a very clean interpretation of transport and is more in line with the concept of leaky containment vessel. The *leakage* is characterized by the relative motion between the tracer isosurface and enclosed air mass. One can think of it as either mass transport in the tracer coordinate or tracer transport in the mass coordinate. This characterization provides a natural conceptual framework for the diagnostic while simplifying the mathematics at the same time. The relative transport of tracer and mass reflects nonconservative (nonadvective) processes because without them the tracer isosurfaces would be a material surface and hence impermeable to mass. This last statement can be labeled as the *nontransport theorem*.

Among other nonconservative processes, we have focused on the horizontal (isentropic) mixing and diabatic circulation. The present formalism allows us to conceptualize and quantify these processes with least ambiguity and assumptions. Above all, use of the mass (area) coordinate has greatly simplified interpretation and diagnosis of horizontal mixing. Most other formalisms (e.g.

TEM; also see Tung 1982, 1986; Plumb and Mahlman 1987) use geographical latitude as the meridional coordinate, and the eddy diffusivity parameterization as a representation of horizontal mixing. However, validity of the eddy diffusivity concept for finite-amplitude eddies in highly inhomogenous environment is at best questionable (McIntyre 1992). Meanwhile, particle/contour advection methods lack useful flux-gradient relationships and require solution of initial value problems to diagnose mixing properties. The present area-equivalent length diagnostic addresses these shortcomings of the previous methods, and makes the link between the chaotic advection and microscale diffusion more transparent: Stirring by chaotic advection stretches the equivalent length, the available interface across which diffusion operates, and hence accelerates mixing of tracer. (See also Yang 1994 and Bowman 1995 for related discussions.) Apart from the *practical* difficulty in its precise measurement, equivalent length identifies well-stirred regions and barriers to mixing *instantaneously*. Formally, the square of equivalent length multiplied by microscale diffusion may be viewed as a Lagrangian equivalence of the eddy diffusion coefficient. Unlike its Eulerian counterpart, however, no parameterization is involved in its definition. It has also been demonstrated that concentrated tracer inhomogeneity, such as edges, can emerge as a result of differential stirring, that is, spatially varying stretching of equivalent lengths. We have also seen that the net effect of mean diabatic circulation is to redistribute mass *within* the container, while horizontal mixing brings about net transport *across* the wall of the container.

More than being a conceptual tool, the MLM diagnostic allows for an efficient processing of the tracer data. A robust mean tracer distribution can be constructed from instantaneous data, without taking time average. Distinction between the polar vortex, surf zone, and tropical reservoir is resolved much more precisely than by the zonal-mean diagnostics because the mass coordinate traces the Lagrangian identity of the wavy edges and hence preserves sharp gradients normal to them. While all diagnostic examples used here have been drawn from high-resolution numerical simulations, in principle the diagnostic can also be applied to the real data with a few caveats. First, the requirement for the data coverage is more stringent since the Lagrangian mean is nonlocal [eqns (7) and (29)]: To compute the mean values at an equivalent latitude (around the wall of the container), one needs information poleward of that latitude (inside the container). Also, horizontal resolution of the data should be sufficiently strong in order to obtain quantitatively meaningful equivalent lengths.

We have developed this formalism with an immediate application to the stratospheric transport in mind, but the diagnostic should also be useful to other transport problems because the formulation is completely general and independent of the underlying dynamics. Indeed, permeability of the tracer edge is a recurring theme in geophysical fluid dynamics. Much parallelism is found over the debates on the roles of the polar vortex edge (McIntyre 1989, Tuck 1989, Schoeberl *et al*. 1992), meandering axis of the Gulf Stream (Bower *et al*. 1985, Halking and Rossby 1985), and midlatitude tropopause (Danielsen 1968, Follows 1992, Appenzeller and Davies 1992, Holton *et al*. 1995) in the tracer and mass transport. The MLM method should lend equally valuable insights into these seemingly unrelated problems if one uses isentropic (isopycnal) co-ordinate in the vertical (Hoskins *et al*. 1985).

The success of the present method has been bought at a price, however. The price, ironically, is the very trick that simplified the formalism—that everything is measured *relative to the containment vessel* without reference to the Eulerian coordinates. The diagnostic provides little information about the position, shape and motion of the container itself. The *equivalent* latitude, for example, can be associated with *geographical* latitude only in the sense of a statistical mean, assuming that the (zonally asymmetric) latitudinal excursion of the tracer contours is sufficiently ergodic. As noted in the introduction, the actual position of the container governs much of the radiative and photochemical processes inside it. This stands as a major hindrance to constructing a two-dimensional chemistry model based on the present formalism. In this regard, Waugh's (1997) geometrical characterization of the size, position and shape of the tracer contours with few ellipses may turn out to be useful.

There are, of course, other unresolved issues. Most fundamental of all, the mass (area)-tracer relation is specific to the tracer. Even if the contours of two tracers enclose the same amount of mass, the Lagrangian identity of the two air masses is different unless the two contours are coincident (Fig.14). Thus, the meaning of m (or φ_e) becomes tracer-specific, and hence it cannot be a common reference for more than one species. Fortunately, this is less of a problem for chemically long-lived tracers which develop similar contour shapes on the isentropic surface. This is because all passive tracers tend to be homogenized streamwise by shear-augmented dispersion (Rhines and Young 1983, see also Plumb and Ko 1992). Hence the contours of two tracers become approximately

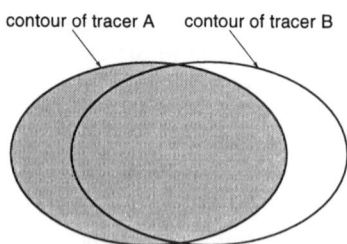

contour of tracer A   contour of tracer B

Figure 14 A schematic of two tracer contours enclosing the same amount of mass.

coincident (although their gradients may differ). Under such circumstance φ_e *is* a common reference for the two tracers, and it can also be shown that the two tracers share the same equivalent length. However, for chemically active species, compact tracer relations generally do not exist [with a few exceptions: for example, a remarkable anticorrelation is observed between the ClO and ozone mixing ratios in the Antarctic air during the development of the ozone hole (Anderson *et al.* 1989)]. This poses another challenge for the modeling of photochemistry using the present formalism. A practical alternative may be to pick a reference tracer and define transport of other species relative to this tracer. Indeed, this was the original theme of MLM thinking (McIntyre 1980). For the modeling purposes, PV may be suitable as such coordinate (e.g. Manney *et al.* 1994; Lary *et al.* 1995), but for diagnostic purposes a quasi-conservative chemical tracer (such as nitrous oxide or methane) may be more practical since PV is not readily computable from the observed stratospheric data.

Another remaining area concerns the effect of vertical diffusion. Although the illustration in section 3 makes it clear that stretching of the tracer isosurface is crucial for efficient mixing, the nature of stretching in the three-dimensional space is more complex. Vertical shear, for example, tends to spread the tracer isosurfaces *horizontally*, and hence make more interface available to *vertical* diffusion. Indeed, based on the fact that vertical shear is typically larger than horizontal shear in the stratosphere, some studies only consider the effect of vertical diffusion (e.g. Prather and Jaffe 1990). This is somewhat misleading, since chaotic advection supports far more efficient mixing than steady shear, and the flow in the winter stratosphere is more chaotic in the horizontal dimensions than in the vertical due to planetary wave breaking. It is thus more reasonable to assume that stretching is largely driven by the horizontal process, but that vertical diffusion becomes important as the generated filaments (sheets) of tracer get tilted over in the presence of vertical shear. [See Hamilton *et al.* 1994 for a related discussion.] In the case of PV, a scale-dependent radiative damping due to optically thick absorbers can in effect act as vertical diffusion, which can prevent the vertical scale of the PV sheets from cascading to small scales comparable to those of the tracer sheets (Haynes and Ward 1993). Incorporating vertical diffusion into the MLM formalism requires knowledge of the local orientation of the tracer isosurface, and the associated geometrical consideration becomes more complicated than the area-length relation of the two-dimensional mixing.

Given these hurdles yet to be cleared, it remains to be seen whether the present method eventually evolves into a fruitful successor of the traditional two-dimensional chemistry models. It seems fair to say at this juncture, however, that as far as the characterization of the leaky containment vessel is concerned, the MLM method offers clean conceptual interpretations and diagnostics, filling the gap between the zonally averaged models and particle advection methods.

Appendix Proof of eqn (8)

Suppose a contour of the tracer q moves slightly in the horizontal plane over the time inclement δt (Fig.15). The displacement can be a result of advection, diffusion or a local source/sink. Initially the contour encloses area A_0, but after δt, it encloses $A_0 + \delta A_2 - \delta A_1$. Assuming q is greater outside the contour,

$$\delta \iint_{q^* \leq q} \chi \, dA \;=\; \iint_{A_0 + \delta A_2 - \delta A_1} \chi(t + \delta t) \, dA \;-\; \iint_{A_0} \chi(t) \, dA$$

$$\approx \iint_{A_0} \delta t \, \frac{\partial \chi}{\partial t} \, dA \;+\; \iint_{\delta A_2} \chi(t) \, dA \;-\; \iint_{\delta A_1} \chi(t) \, dA$$

$$= \iint_{q^* \leq q} \delta t \, \frac{\partial \chi}{\partial t} \, dA \;-\; \oint_q \chi \frac{\partial q / \partial t}{|\nabla q|} \, \delta t \, dl$$

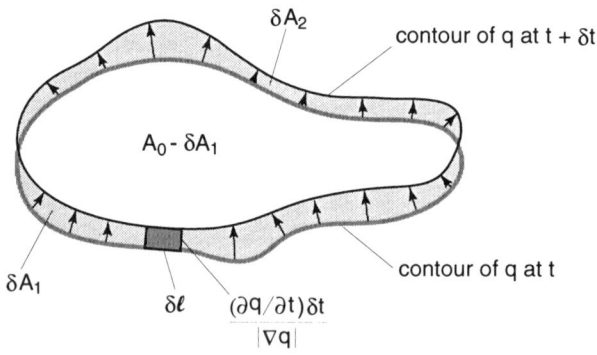

Figure 15 Horizontal displacement of a closed contour of tracer q. Here we assume that q is smaller inside the contour. Both enclosed area and shape of the contour can change during the displacement.

$$= \delta t \left(\iint\limits_{q^* \le q} \frac{\partial \chi}{\partial t} \, dA \; - \; \frac{\partial}{\partial q} \iint\limits_{q^* \le q} \chi \frac{\partial q}{\partial t} \, dA \right), \quad \leftarrow Eqns \; (6),(4)$$

where χ is an arbitrary scalar. From the second line to the third above, we used the fact that the area element on δA_1 and δA_2 segments is given by $\pm (\partial q / \partial t) |\nabla q|^{-1} \delta l \, \delta t$ and that $\partial q / \partial t$ changes sign between the two segments. Dividing both sides by δt and taking $\delta t \to 0$, one obtains eqn (8). The result also holds for the opposite gradient of q and can be generalized for more complicated displacement patterns.

Acknowledgments

Of the many individuals I am indebted to, I would especially like to thank Mike Wallace and the late Stephen Fels who served, in many ways, as gateways to my inquiry of mixing and stratosphere, respectively. Jerry Mahlman provided GFDL's resources, SKYHI output and many helpful discussions. I have learned much about the stratosphere and transport through discussions/correspondence with the following group of people to whom my gratitude is due: A. Plumb, J. Holton, R. Pierrehumbert, P. Haynes, T. Shepherd, M. McIntyre, D. Waugh, J. Scinocca, H. Yang, S. Strahan, J. Eluszkiewicz, J. Abbatt, J. Frederick, L. Mickley, G. Holloway, P. Rhines, W. Young, W. Grose, K. Hamilton, K. Bowman, M. Geller, W. Robinson. I am also grateful to J. Wilson, operational staff of GFDL, J. Valdes and J. Ma for their technical assistance. This research has been jointly supported by National Science Foundation Grant ATM-9400574, National Astronautics and Space Administration Grants NAGW-4281 and NAG5-2792.

References

Anderson, J. G., W. H. Brune, S. A. Lloyd, D. W. Toohey, S. P. Sander, W. L. Starr, M. Loewenstein and J. R. Podolske, 1989: Kinetics of ozone destruction by ClO and BrO within the Antarctic vortex: An analysis based on in situ ER-2 data. *J. Geophys. Res.* **94**, 11480-11520.

Andrews, D. G., J. R. Holton and C. B. Leovy, 1987: *Middle Atmosphere Dynamics*. Academic Press, 489pp.

Andrews, D. G. and M. E. McIntyre, 1976: Planetary waves in horizontal and vertical shear: The generalized Eliassen-Palm relation and the mean zonal acceleration. *J. Atmos. Sci.* **33**, 2031-2048.

Andrews, D. G. and M. E. McIntyre, 1978: An exact theory of nonlinear waves on a Lagrangian-mean flow. *J. Fluid Mech.* **89**, 609-646.

Appenzeller, C. and H. C. Davies, 1992: Structure of stratospheric intrusions into the troposphere. *Nature* **358**, 570-572.

Baldwin, M. P. and J. R. Holton, 1988: Climatology of the stratospheric polar vortex and planetary wave breaking. *J. Atmos. Sci.* **45**, 1123-1142.

Baldwin, M. P. and K. K. Tung, 1994: Extra-tropical QBO signals in angular momentum and wave forcing. *Geophys. Res. Lett.* **21**, 2717-2720.

Batchelor, G. K., 1959: Small-scale variation of convected quantities like temperature in turbulent fluid—Part I: General discussion and the case of small conductivity. *J. Fluid Mech.* **5**, 113-133.

Bower, A. S., H. T. Rossby and J. L. Lilibridge, 1985: The Gulf Stream—barrier or blender? *J. Phys. Ocean.* **15**, 24-32.

Bowman, K. P., 1993: Large-scale isentropic mixing properties of the Antarctic polar vortex from analyzed winds. *J. Geophys. Res.* **98**, 23013-23017.

Bowman, K. P., 1995: Diffusive transport by breaking waves. *J. Atmos. Sci.* **52**, 2416-2427.

Boyd, J. P., 1976: The noninteraction of waves with the zonally averaged flow on a spherical earth and the interrelationships of eddy fluxes of energy, heat and momentum. *J. Atmos. Sci.* **33**, 2285-2291.

Brewer, A. W., 1949: Evidence for a world circulation provided by the measurements of the helium and water vapor distribution in the stratosphere. *Quart. J. Roy. Meteor. Soc.* **75**, 351-363.

Butchart, N. and E. E. Remsberg, 1986: The area of the stratospheric polar vortex as a diagnostic for tracer transport on an isentropic surface. *J. Atmos. Sci.* **43**, 1319-1339.

Charney, J. G. and P. G. Drazin, 1961: Propagation of planetary-scale disturbances from the lower into the upper atmosphere. *J. Geophys. Res.* **66**, 83-110.

Chen, P., 1994: The permeability of the Antarctic vortex edge. *J. Geophys. Res.* **99**, 20563-20571.

Dahlberg, S. P. and K. P. Bowman, 1994: Climatology of large-scale isentropic mixing in the Arctic winter stratosphere from analyzed winds. *J. Geophys. Res.* **99**, 20585-20599.

Danielsen, E. F., 1968: Stratospheric-tropospheric exchange based on radioactivity, ozone, and potential vorticity. *J. Atmos. Sci.* **25**, 502-518.

Dobson, G. M. B. (1956). Origin and distribution of the polyatomic molecules in the atmosphere. Proc. R. Soc., London, **A256**, 187-193.

Drazin, P. G. and W. H. Reid, 1981: *Hydrodynamic stability*. Cambridge University Press, Cambridge, 527pp.

Dritschel, D. G., 1988: The repeated filamentation of two-dimensional vorticity interfaces. *J. Fluid Mech.* **194**, 511-547.

Dunkerton, T. J., 1978: On the mean meridional mass motions of the stratosphere and mesosphere. *J. Atmos. Sci.* **35**, 2325-2333.

Dunkerton, T. J. and D. P. Delisi, 1986: Evolution of potential vorticity in the winter stratosphere of January-February 1979. *J. Geophys. Res.* **91**, 1199-1208.

Dunkerton, T. J., C.-P. F. Hsu and M. E. McIntyre, 1981: Some Eulerian and Lagrangian diagnostics for a model stratospheric warming. *J. Atmos. Sci.* **38**, 819-843.

Eluszkiewicz, J., D. Crisp, R. Zurek, L. Elson, E. Fishbein, L. Froidevaux, J. Waters, R. G. Grainger, A. Lambert, R. Harwood and G. Peckham, 1996: Residual circulation in the stratosphere and lower mesosphere as diagnosed from Microwave Limb Sounder data. *J. Atmos. Sci.* **53**, 217-240.

Eluszkiewicz, J., R. A. Plumb and N. Nakamura, 1995: Dynamics of wintertime stratospheric transport in the Geophysical Fluid Dynamics Laboratory "SKYHI" general circulation model. *J. Geophys. Res.* **100**, 20883-20900.

Fels, S. B., 1985: Radiative-dynamical interactions in the middle atmosphere. In *Advances in Geophysics*. S. Manabe, Ed. Academic Press. Vol **A**, 277-300.

Fels, S. B., J. D. Mahlman, M. D. Schwarzkopf and R. W. Sinclair, 1980: Stratospheric sensitivity to perturbations in ozone and carbon dioxide: radiative and dynamical response. *J. Atmos. Sci.* **37**, 2265-2297.

Follows, M. J., 1992: On the cross-tropopause exchange of air. *J. Atmos. Sci.* **49**, 879-882.

Garcia, R. R., 1991: Parameterization of planetary wave breaking in the middle atmosphere. *J. Atmos. Sci.* **48**, 1405-1419.

Geller, M. A., E. R. Nash, M. F. Wu and J. A. Rosenfield, 1992: Residual circulations calculated from satellite data: Their relations to observed temperature and ozone distributions. *J. Atmos. Sci.* **49**, 1127-1137.

Gille, J. C., L. V. Lyjak and A. K. Smith, 1987: The global residual mean circulation in the middle atmosphere for the northern winter period. *J. Atmos. Sci.* **44**, 1437-1452.

Halking, D. and T. Rossby, 1985: The structure and transport of the Gulf Stream at 73W. *J. Phys. Ocean.* **15**, 1439-1452.

Hamilton, K. P., 1995: Interannual variability in the Northern Hemisphere winter middle atmosphere in control and perturbed experiments with the SKYHI general circulation model. *J. Atmos. Sci.* **52**, 44-66.

Hamilton, K. P., R. J. Wilson, J. D. Mahlman and L. J. Umscheid, 1995: Climatology of the SKYHI troposphere-stratosphere-mesosphere general circulation model. *J. Atmos. Sci.* **52**, 5-43.

Hamilton, K. P., R. J. Wilson and H. Vahlenkamp, 1994: Three-dimensional visualization of the polar stratospheric vortex. *Canadian Met. and Ocean. Soc. Bulletin* **22**(#4), 4-6.

Haynes, P. H., 1985: Nonlinear instability of a Rossby-wave critical layer. *J. Fluid Mech.* **161**, 493-511.

Haynes, P. H. and W. E. Ward, 1993: The effect of realistic radiative transfer on potential vorticity structures, including the influence of background shear and strain. *J. Atmos. Sci.* **50**, 3431-3453.

Hitchman, M. H. and C. B. Leovy, 1986: Evolution of the zonal mean state in the equatorial middle atmosphere during October 1978-May 1979. *J. Atmos. Sci.* **43**, 3159-3176.

Holton, J. R., 1986: Meridional distribution of stratospheric trace constituents. *J. Atmos. Sci.* **43**, 1238-1242.

Holton, J. R. and J. Austin, 1991: The influence of the equatorial QBO on sudden stratospheric warmings. *J. Atmos. Sci.* **48**, 607-618.

Holton, J. R., P. H. Haynes, M. E. McIntyre, A. R. Douglass, R. B. Rood and L. Pfister, 1995: Stratosphere-troposphere exchange. *Rev. Geophys.* **33**(4), 403-439.

Hoskins, B. J., 1991: Towards a PV-theta view of the general circulation. *Tellus* **43AB**, 27-35.

Hoskins, B. J., M. E. McIntyre and A. W. Robertson, 1985: On the use and significance of isentropic potential vorticity maps. *Quart. J. Roy. Meteor. Soc.* **111**, 877-946.

Juckes, M. N. and M. E. McIntyre, 1987: A high-resolution, one-layer model of breaking planetary waves in the stratosphere. Nature. **328:** 590-596.

Kida, H., 1983: General circulation of air parcels and transport characteristics derived from a hemispheric GCM. Part 2: Very long-term motions of air parcels in the troposphere and stratosphere. *J. Meteor. Soc. Japan* **61**, 510-523.

Killworth, P. D. and M. E. McIntyre, 1985: Do Rossby-wave critical layers absorb, reflect, or over-reflect? *J. Fluid Mech.* **161**, 449-492.

Lait, L. R., 1994: An alternative form for potential vorticity. *J. Atmos. Sci.* **51**, 1754-1759.

Lary, D. J., M. P. Chipperfield, J. A. Pyle, W. A. Norton and L. P. Riishojgaad, 1995: Three dimensional tracer initialisation and general diagnostics using equivalent PV latitude - potential temperature coordinates. *Quart. J. Roy. Meteor. Soc.* **121**, 187-210.

Mahlman, J. D., I. Levy H. and W. J. Moxim, 1986: Three-dimensional simulations of stratospheric N_2O: predictions for other trace constituents. *J. Geophys. Res.* **91**, 2687-2707.

Mahlman, J. D. and L. J. Umscheid, 1987: Comprehensive modeling of the middle atmosphere: The influence of horizontal resolution. In *Transport Pro-*

cesses in the Middle Atmosphere. G. Visconti and R. Garcia, Ed. Hingham, Mass., D. Reildel.

Manney, G. L., R. W. Zurek, A. O'Neal, and R. Swinbank, 1994: On the motion of air through the stratospheric polar vortex. *J. Atmos. Sci.* **51**, 2973-2994.

Matsuno, T., 1971: A dynamical model of the stratospheric sudden warming. *J. Atmos. Sci.* **28**, 1479-1494.

McIntyre, M. E., 1980: Towards a Lagrangian-mean description of stratospheric circulations and chemical transport. *Phil. Trans. Roy. Soc.*, London, **A296**, 129-148.

McIntyre, M. E., 1982: How well do we understand the dynamics of stratospheric warmings? *J. Meteor. Soc. Japan* **60**, 37-65.

McIntyre, M. E., 1989: On the Antarctic Ozone Hole. *J. Atmos. Terr. Phys.* **51**, 29-43.

McIntyre, M. E. , 1992: Atmospheric dynamics: Some fundamentals, with observational implications. *Int. School Phys. "Enrico Fermi" CXV Course*, North-Holland, 313-386.

McIntyre, M. E. and T. N. Palmer, 1983: Breaking planetary waves in the stratosphere. *Nature* **305**, 593-600.

McIntyre, M. E. and T. N. Palmer, 1984: The "surf zone" in the stratosphere. *J. Atmos. Terr. Phys.* **46**, 825-849.

Nakamura, M. and R. A. Plumb, 1994: The effects of flow asymmetry on the direction of Rossby wave breaking. *J. Atmos. Sci.* **51**, 2031-2045.

Nakamura, N., 1995: Modified Lagrangian-mean diagnostics of the stratospheric polar vortices. Part I. Formulation and analysis of GFDL SKYHI GCM. *J. Atmos. Sci.* **52**, 2096-2108.

Nakamura, N., 1996: Two-dimensional mixing, edge formation, and permeability diagnosed in an area coordinate. *J. Atmos. Sci.* **53**, 1524-1537.

Nakamura, N. and J. Ma, 1997: Modified Lagrangian-mean diagnostics of the stratospheric polar vortices: Part II. Nitrous oxide and seasonal barrier migration in Cryogenic Etalon Limb Array Spectrometer measurements and SKYHI general circulation model. *J. Geophys. Res.* **101**, revised.

Nash, E. R., P. A. Newman, J. E. Rosenfield and M. R. Schoeberl, 1996: An objective determination of the polar vortex using Ertel's potential vorticity. *J. Geophys. Res.* **101**, 9471-9478.

Newman, P. A., L. R. Lait, M. R. Schoeberl, E. R. Nash, K. Kelly, D. Fahey, R. M. Nagatani, D. Toohey, L. Avallone and J. Anderson, 1993: Stratospheric meteorological conditions in the Arctic polar vortex, 1991 to 1992. *Science* **261**, 1134-1136.

Norton, W. A., 1994: Breaking Rossby waves in a model stratosphere diagnosed by a vortex-following coordinate system and a technique for advecting material contours. *J. Atmos. Sci.* **51**, 654-673.

Ottino, J. M., 1989: *The kinematics of mixing: stretching, chaos, and transport.* Cambridge University Press, 364pp.

Pierce, R. B. and T. D. Fairlie, 1993: Chaotic advection in the stratosphere: Implications for the dispersal of chemically perturbed air from the polar vortex. *J. Geophys. Res.* **98**, 18589-18595.

Pierce, R. B., T. D. Fairlie, W. L. Grose, R. Swinbank and A. O'Neil, 1994: Mixing processes within the polar night jet. *J. Atmos. Sci.* **51**, 2957-2972.

Pierrehumbert, R. T., 1991a: Large-scale horizontal mixing in planetary atmospheres. *Phys. Fluids* **3A**, 1250-1260.

Pierrehumbert, R. T., 1991b: Chaotic mixing of tracer and vorticity by modulated travelling Rossby waves. *Geophys. Astrophys. Fluid Dyn.* **58**, 285-319.

Pierrehumbert, R. T., 1994: Tracer microstructure in the large-eddy dominated regime. *Chaos, Solitons & Fractals* **4**, 1091-1110.

Pierrehumbert, R. T. and H. Yang, 1993: Global chaotic mixing on isentropic surfaces. *J. Atmos. Sci.* **50**, 2462-2480.

Plumb, R. A., 1996: A "tropical pipe" model of stratospheric transport. *J. Geophys. Res.* **101**, 3957-3972.

Plumb, R. A. and M. K. W. Ko, 1992: Interrelationships between mixing ratios of long-lived stratospheric constituents. *J. Geophys. Res.* **97**, 10145-10156.

Plumb, R. A. and J. D. Mahlman, 1987: The zonally-averaged transport characteristics of the GFDL general circulation/transport model. *J. Atmos. Sci.* **44**, 262-270.

Plumb, R. A., D. W. Waugh, R. J. Atkinson, P. A. Newman, L. R. Lait, M. R. Schoeberl, E. V. Browel, A. J. Simmons and M. Loewenstein, 1994: Intrusion into the lower stratospheric Arctic vortex during the winter of 1991-1992. *J. Geophys. Res.* **99**, 1089-1105.

Polvani, L. M. and R. A. Plumb, 1992: Rossby wave breaking, microbreaking, filamentation, and secondary vortex formation: the dynamics of a perturbed vortex. *J. Atmos. Sci.* **49**, 462-476.

Polvani, L. M., D. W. Waugh and R. A. Plumb, 1995: On the subtropical edge of the stratospheric surf zone. *J. Atmos. Sci.* **52**, 1288-1309.

Prather, M. and A. H. Jaffe, 1990: Global impact of the Antarctic ozone hole: chemical propagation. *J. Geophys. Res.* **95**, 3473-3492.

Randel, W. J., J. C. Gille, A. E. Roche, J. B. Kumer, J. L. Mergenthaler, J. W. Waters, E. F. Fishbein and W. A. Lahoz, 1993: Planetary wave mixing in the subtropical stratosphere observed in UARS constituent data. *Nature* **365**, 533-535.

Rhines, P. B. and W. R. Young, 1983: How rapidly is a passive scalar mixed within closed streamlines? *J. Fluid Mech.* **133**, 133-145.

Roche, A. E., J. B. Kumer, R. W. Nightingale, J. L. Mergenthaler, G. A. Ely, P. L. Bailey, S. T. Massie, J. C. Gille, D. P. Edwards, M. R. Gunson, M. C. Abrams, G. C. Toon, C. R. Webster, W. A. Traub, K. W. Jucks, D. G. Johnson, D. G. Murcray, F. H. Marcray, A. Goldman and E. C. Zipf, 1996: Validation of CH_4 and N_2O measurements by the cryogenic limb array etalon spectrometer instrument on the Upper Atmosphere Research Satellite. *J. Geophys. Res.* **101**, 9679-9710.

Rosenfield, J. E., P. A. Newman and M. R. Schoeberl, 1994: Computations of diabatic descent in the stratospheric polar vortex. *J. Geophys. Res.* **99**, 16677-16689.

Ruth, S. L., J. J. Remedios, B. N. Lawrence and F. W. Taylor, 1994: Measurements of N_2O by the UARS Improved Stratospheric and Mesospheric Sounder during the early northern winter 1991/92. *J. Atmos. Sci.* **51**, 2818-2833.

Schoeberl, M. R. and L. R. Lait, 1991: Conservative coordinate transformations for atmospheric measurements. In *EOS NATO Summer School.* G. Visconti and J. Gille, Ed. American Geophysical Union.

Schoeberl, M. R., L. R. Lait, P. A. Newman and J. E. Rosenfield, 1992: The structure of the polar vortex. *J. Geophys. Res.* **97**, 7859-7882.

Scinocca, J. F., 1995: The mixing of mass and momentum by Kelvin-Helmholtz billows. *J. Atmos. Sci.* **52**, 2509-2530.

Solomon, S., J. P. Smith, R. W. Sanders, L. Perliski, H. L. Miller, G. H. Mount, J. G. Keys and A. L. Schmeltekopf, 1993: Visible and near-ultraviolet spectroscopy at McMurdo Station, Antarctica, 8. Observations of nighttime NO_2 and NO_3 from April to October 1991. *J. Geophys. Res.* **98**, 993-1000.

Strahan, S. E. and J. D. Mahlman, 1994a: Evaluation of the SKYHI general circulation model using aircraft N_2O measurements. 1. Polar winter stratospheric meteorology and tracer morphology. *J. Geophys. Res.* **99**(D5), 10305-10318.

Strahan, S. E. and J. D. Mahlman, 1994b: Evaluation of the SKYHI general circulation model using aircraft N_2O measurements. 2. Tracer variability and diabatic meridional circulation. *J. Geophys. Res.* **99**, 10319-10332.

Sutton, R. T., H. MacLean, R. Swinbank, A. O'Neil and F. W. Taylor, 1994: High-resolution stratospheric tracer fields estimated from satellite observations using Lagrangian trajectory calculations. *J. Atmos. Sci.* **51**, 2995-3005.

Tao, X. and A. F. Tuck, 1994: On the distribution of cold air near the vortex edge in the lower stratosphere. *J. Geophys. Res.* **99**, 3431-3450.

Tuck, A. F., 1989: Synoptic and chemical evolution of the Antarctic vortex in late winter and early spring, 1987. *J. Geophys. Res.* **94**, 11687-11737.

Tung, K. K., 1982: On the two-dimensional transport of stratospheric trace gases in isentropic coordinates. *J. Atmos. Sci.* **39**, 2230-2355.

Tung, K. K., 1986: Nongeostrophic theory of zonally averaged circulation. Part I: Formulation. *J. Atmos. Sci.* **43**, 2600-2618.

Waugh, D. W., 1997: Elliptical diagnostics of the stratospheric polar vortices. *Quart. J. Roy. Meteor. Soc.* **123**, in press.

Waugh, D. W. and R. A. Plumb, 1994: Contour advection with surgery: A technique for investigating fine scale structure in tracer transport. *J. Atmos. Sci.* **51**, 530-540.

Waugh, D. W., R. A. Plumb, R. J. Atkinson, M. R. Schoeberl, L. R. Lait, P. A. Newman, M. Loewenstein, D. W. Toohey, L. M. Avallone, C. R. Webster and R. D. May, 1994: Transport out of the lower stratospheric Arctic vortex by Rossby wave breaking. *J. Geophys. Res.* **99**, 1071-1088.

Weiss, J. B., 1991: Transport and mixing in travelling waves. *Phys. Fluids* **A3**, 1379-1384.

Welander, P., 1955: Studies on the general development of motion in a two-dimensional ideal fluid. *Tellus* **7**, 141-156.

Wiggins, S., 1988: Stirred but not mixed. *Nature* **333**, 395-396.

Yang, H., 1993: Chaotic mixing and transport in wave systems and the atmosphere. *Inter. J. Bifurc. Chaos* **3**, 1423-1445.

Yang, H., 1994: On the relative importance between chaotic mixing and diffusion. *Phys. Lett. A* **185**, 191-195.

Yang, H., K. K. Tung and E. Olaguer, 1990: Nongeostrophic theory of zonally averaged circulation. Part II: Eliassen-Palm flux divergence and isentropic mixing coefficient. *J. Atmos. Sci.* **47**, 215-241.

Computational Mechanics Publications

Nonlinear Ocean Waves

Edited by: W. PERRIE, Bedford Institute of Oceanography, Canada

Ocean waves are generated and evolve in space and time, sometimes propagating over thousands of kilometres and growing to several 10's of meters in height. Understanding these waves involves looking at the processes that drive them and determine their development including the energy removed from waves by wave breaking and white-capping, and non-linear wave-wave interactions. In this study we consider (i) how observed waves grow and develop, maintaining and equilibrium with the wind, being driven by wind and also modifying the wind, (ii) how ideal potential waves grow and develop, as well as the spectra of wind-wave turbulence, and (iii) modelling of non-linear wave-wave interactions, wind input and wave dissipation in shallow water and turning wind situations.

Partial contents: Wind-forced strong wave interactions and quasi-local equilibrium between wind and windsea with the friction velocity proportionality; Mathematics and approximation of the nonlinear wave-wave interactions, Relating nonlinear energy cascades to wind input and wave breaking dissipation; Turbulence of capillary waves-theory and numerical simulation.

Series: Advances in Fluid Mechanics, Vol 17
ISBN: 185312 4141 Nov 1997
aps 300 pp apx £88.00/$135.00

Free Surface Flows with Viscocity

Edited by P. A. TYVAND, *Agricultural University of Norway, Aas, Norway*

Free surface flows including fluid viscosity is a challenging field of research. Its development faces the fact that inviscid theory is now reaching a mature stage. This monograph gives a representative picture of the current basic research on viscous free surface flows. New analytical techniques are presented, partly in combination with experiments Reviews are

give of the established literature on viscous ocean waves and fluid films.

Partial contents: Water waves; Ship waves; Ocean waves on a rotating earth; Stokes drift; Wave damping; Vorticity near a free surface; Internal waves; and viscous thin-layer flows.

Series: Advances in Fluid Mechanics, Vol 16
ISBN: 1853122955 1998, 280pp, £84.00/ £130.00

Self-Sustaining Mechanisms of Wall Turbulence

Edited by: R.L. PANTON, *University of Texas, USA*

This book unravels some of the mysteries surrounding wall turbulence and their ability to be self-sustaining. As the title of this book suggests, the use of the plural "mechanisms" implies that there may be a number of reasons why self-sustainment is possible. The authors of this book were encouraged to discuss not only what they know but to speculate about ideas that require numerical or experimental testing and verification, thus paving the way for future research.

Partial Contents: A Brief History of Boundary Layer Structure Research; The Role of Wall Vortices in Producing Turbulence; A View of the Structure of Turbulence Boundary Layers; Reynolds Stress Producing Motions in Smooth and Rough Wall Boundary Layers; Genesis and Dynamics of Coherent Structures in Near-Wall Turbulence: A New Look.

Series: *Advances in Fluid Mechanics, Vol 15*
ISBN: 1853124532 1997 425pp £84.00/ $134.00

All prices correct at time of going to press. All books are available from your bookseller or in case of difficulty direct from the Publisher.

Computational Mechanics Publications
Ashurst Lodge, Ashurst, Southampton,
SO40 7AA, UK.
Tel: 44 (0)1703 293223
Fax: 44 (0)1703292853
Email: cmp@cmp.co.uk

Computational Mechanics Publications

Laminar and Turbulent Boundary Layers

Edited by: M. RAHMAN, *Technical University of Nova Scotia, Canada*
This volume contains six state-of-the-art chapters on laminar and turbulent boundary layers, presenting a comprehensive picture of the use of laminar and turbulence boundary layers in solving problems of a physical nature. Topics covered include the propagation of weak discontinuities wiht applications to fluid mechanics; solutionns and anomalies of the boundary layer equations for axial flow along a cirucalr cylinder and wall pressure fluctuations under trubulent boundary layers.
Partial Contents: Some Aspects of Perturbation Solutions Arising in 2D Laminar Boundary Layers; Weak Discontinuities and Rays in Hyperbolic Systems - with Applications; Boundary Layer Flow along a Circular Cylinder.
Series: *Advances in Fluid Mechanics, Vol 14*
ISBN: 1853122947 1997 192pp £64.00/ $99.00

Fluid Transport in Porous Media

Edited by: P. DU PLESSIS, *University of Stellenbosch, South Africa*
This book is devoted to advances in mathematical modelling and to the application of fluid mechanics to fluid transport and electric conduction in porous media. The theoretical work presented looks at porous media in a general sense, i.e. the results given are aimed at general applications to a wide variety of porous medium microstructures and length scales. It also forms part of a concerted effort to enhance mutual interactive research among mathematical, computational and experimental modellers and practitioners in the field.
Partial Contents: Pore-Scale Modelling of Interstitial Transport Phenomena; Recent Advances in Theories of Two-Phase Flow in Porous Media; Flow in Rotating Porous Media.
Series: *Advances in Fluid Mechanics, Vol 13*
ISBN: 185312429X 1997 320pp £98.00/ $147.00

Nonlinear Instability Analysis

Edited by: L. DEBNATH and S. R. CHOUDHURY, *University of Central Florida, USA*
This monograph presents a look at state-of-the-art developments in nonlinear instability analysis. It consists of eight chapters written by leading researchers in the field and provides a invaluable guide to modern mathematical techniques and research literature.
Series: *Advances in Fluid Mechanics, Vol 12*
ISBN: 1853124281 1997 336pp £108.00/ $169.00

Flows at Large Reynolds Numbers

Edited by: H. SCHMITT, *Institute of Fluid Mechanics, Deutsche Forschungsanstalt, fur Luft- and Raumfahrt (DLR), Gottingen, Germany*
Many types of flow of air, water, or other fluids, which are of technical interest, occur at large Reynolds numbers. This book presents the state-of-the-art in the calculation of incompressible and compressible flows at large Reynolds numbers, providing fascinating reading for any scientist or engineer wishing for enlightenment into the use of numerical methods in calculating Large Reynolds numbers.
Series: *Advances in Fluid Mechanics, Vol 11*
ISBN: 1853123838 1997 424pp £118.00/$188.00

Computational Mechanics Publications

Gravity Waves in Water of Finite Depth

Edited by: **J. HUNT,** *Reading University, UK*

This book looks at how numerical computation and new analytical techniques are being used to understand the process by which ocean waves are transformed as they advance towards the shore. Attention is focused on the nearshore region, seaward of the surf zone.

Partial Contents: Gravity Waves in Water of Finite Depth - A General Introduction; Linear Wave Scattering by Two-Dimensional Topography; Nonlinear, Dispersive Long Waves in Water of Variable Depth; Parabolic Modelling of Water Waves; Oscillatory Flow Over Rippled Beds.

Series: *Advances in Fluid Mechanics, Vol10*
ISBN: 185312351X   1996   400pp
£145.00/$217.00

Advances in Fluid Mechanics

Edited by: **M. RAHMAN,** *Technical University of Nova Scotia, Canada* and **C.A. BREBBIA,** *Wessex Institute of Technology, UK*

Fluid mechanics has become an area of academic and industrial research, not only for fundamental work, but in the solution of practical engineering problems. The dramatic increase in computer power has enabled computational techniques to be applied more successfully to problems associated with fluid mechanics. This book provides a unique guide to research in this expanding field, with papers written by a number of key individuals active in the implementation of numerical techniques in fluid mechanics.

Partial Contents: Geophysical Fluid Dynamics; Environmental Fluid Mechanics; Channel and River Flow; Coastal Sea Modelling and Data Assimilation in Oceanic Models; Non-Linear Ocean Waves: Traveling Waves in in Multi-Phase Flows; Multi-Phase Flow; Turbulence.

Series: *Advances in Fluid Mechanics, Vol 9*
ISBN: 1853124524   1996   400pp
£145.00/$217.00

Mathematical Techniques for Water Waves

Edited by: **B.N. MANDAL,** *Indian Statistical Institute, Calcutta, India*

The topic of water waves is an important branch of fluid mechanics. This monograph looks at the use of mathematical techniques in solving water wave problems. It consists of ten chapters, each one written by a leading authority in the field.

Partial Contents: Complementary Methods for Scattering by Thin Barriers; The Use of Multipoles in Channel Problems; Analytical Dynamics of Wave-Body Interactions; The Use of Green's Theorem in Water Wave Problems; Interaction of Water Waves with Thin Plates;

Series: *Advances in Fluid Mechanics, Vol 8*
ISBN: 1853124133   1997   368pp   £108.00/$168.00

Sedimentation of Small Particles in a Viscous Fluid

Edited by: **E.M. TORY,** *Mount Allison University, Canada*

This book contains eight chapters written by leading researchers in the field. The objective is to present a unified treatment of the subject rather than a disjointed collection of articles. This volume therefore concentrates on the sedimentation of particles which are small enough that inertial and unsteady effects can be neglected, but large enough that Brownian motion is negligible.

Series: *Advances in Fluid Mechanics, Vol 7*
ISBN: 1853123579; 1562522809 (US, Canada, Mexico) 1996   304pp   £88.00/$135.00

All prices correct at time of going to press. All books are available from your bookseller or in case of difficulty direct from the Publisher.

Computational Mechanics Publications
Ashurst Lodge, Ashurst, Southampton, SO40 7AA, UK.
Tel: 44 (0)1703 293223
Fax: 44 (0)1703 292853
Email: cmp@cmp.co.uk

Computational Mechanics Publications

Potential Flow of Fluids

Edited by: **M. RAHMAN**, *Technical University of Nova Scotia, Canada*
This book is a compilation of papers covering a selection of advanced topics concerning the Potential Flow of Fluids. It presents 7 chapters logically arranged to present a comprehensive account of developments in this important subject area.
Partial Contents: Some Recent Advances on Wave Effects on Large Offshore Structures; Prediction of Wave Breaking Processes at the Coastline; Wave Breaking Simulation.
Series: *Advances in Fluid Mechanics, Vol 6*
ISBN: 1853123560; 1562522795 (US, Canada, Mexico) 1995 264pp £79.00/ $121.00

Advances in Marine Hydrodynamics

Edited by: **M. OHKUSU**, *Kyushu University, Japan*
The aim of this book is to provide a comprehensive review of advances in marine hydrodynamics, with particular emphasis on their theoretical methods and numerical implementation.
Partial Contents: Ship Resistance and Flow Computations; Hydrodynamics of Ships in Waves; Hydrodynamics of High Speed Vehicles; Computation of Wave Ship Interactions; Cavitation; Theory and Numerical Methods for the Hydrodynamic Analysis of Marine Propulsors.
Series: *Advances in Fluid Mechanics, Vol 5*
ISBN: 1853122874; 1562522116 (US, Canada, Mexico) 1996 384pp £112.00/ $170.00

B E Applications in Fluid Mechanics

Edited by: **H. POWER**, *Wessex Institute of Technology, Southampton, UK*
The Boundary Element Method (BEM) is now a well established numerical technique for the analysis of engineering problems, particular for those involving linear analysis.

One of the main advantages in using this method, is that it drastically reduces the amount of data preparation needed. By applying this method to the field of fluid dynamics, scientists have been able to deal with a number of complex problems, making this one of the most challenging areas of computational mechanics to-date.
Partial Contents: Boundary Element Approach to Laplacian Moving Boundary Problems; Recent Advances in the BEM Modeling of Nonlinear Water Waves; Transonic Field-Boundary Element Computations; Stokes Flow in the Presence of Interfaces; Simulation of Viscous Sintering; CDL-IEM for the Solution of the 2-D Navir-Stokes Equations at Small Reynolds Number: via Singular Perturbation Technique.
Series: *Advances in Fluid Mechanics, Vol 4*
ISBN: 1853122882; 1562522124 (US, Canada, Mexico) 1995 376pp £108.00/$167.00

Fluid Structure Interaction in Offshore Engineering

Edited by: **S.K. CHAKRABARTI**, *Chicago Bridge & Iron Company, Illinois, USA*
Series: *Advances in Fluid Mechanics, Vol 1*
ISBN:1853122807; 1562522043 (US, Canada, Mexico) 1994 256pp £76.00/ $117.00

Ocean Waves Engineering

Edited by: **M. RAHMAN**, *Technical University of Nova Scotia, Canada*
Series: *Advances in Fluid Mechanics, Vol 2*
ISBN: 1853122858; 1562522094 1994 240pp £76.00/$117.00

All prices correct at time of going to press. All books are available from your bookseller or in case of difficulty direct from the Publisher.

Computational Mechanics Publications
Ashurst Lodge, Ashurst, Southampton,
SO40 7AA, UK.
Tel: 44 (0)1703 293223
Fax: 44 (0)1703 292853
Email: cmp@cmp.co.uk